VIRUSES AND BACTERIA

Hands-On, Minds-On Investigations for Middle to High School Students

Dennis Holley

CRITICAL THINKING BOOKS & SOFTWARE
www.criticalthinking.com
P.O. Box 448 • Pacific Grove • CA 93950-0448
Phone 800-458-4849 • FAX 831-393-3277
ISBN 0-89455-717-3
Printed in the United States of America

This book is dedicated to

my parents, for encouraging me to wonder and explore;
my wife, for her patience and understanding;
my students, for teaching me more than they will ever know.

Table of Contents

Introduction

Beyond the limits of human sight exists another world, parallel to our own. This microbial universe abounds with strange beauty and puzzling mysteries. Inhabiting this microcosmos are creatures as strange and bizarre as any imaginable aliens from distant star systems.

The term *microbe* is used in biology to denote any tiny living creature or single-celled organism. In this book, microbe refers only to those submicroscopic nonliving particles called viruses and to those microscopic living cells we know as bacteria.

Despite their tiny size (or perhaps because of it), microbes are truly the dominant life form

on earth. In their antiquity and abundance they eclipse all other life forms. Human skin alone harbors some 100,000 bacteria per square centimeter. From a microbe's point of view, you rival the thickest rain forest of the Amazon or the lushest coral reef in the Pacific Ocean. All ecological systems, and thus life itself, would collapse without microbes.

May this book enlighten you to the marvels of the hidden world of microbes.

I study the lives on a leaf: the little
Sleepers, numb nudgers in cold dimensions,
Beetles in caves, newts, stone-deaf fishes,
Lice tethered to long limp subterranean
weeds,
Squirmers in bogs,
And bacterial creepers.

— Theodore Roethke

How to Use This Book

Who is this guy?

The author is an active classroom teacher. As such, I know and face the same challenges as you do on a daily basis. I also know that you want understandable background information and workable lab activities. I have tried to give you both in this book. We have kept the formidable biology terminology to a minimum and included low-tech, relatively inexpensive lab activities that will bring the information and concepts to life.

What is the purpose of this book?

This book was written to be integrated into and supplement your present curriculum. The intention is not to have you create a microbiology-type mini-course or to throw out what you are doing now. Rather, I hope that you will use the information and activities contained herein to enliven and enhance your current materials on microbes.

Will this book fit my situation?

With slight variations here and there, the information and explorations in this book will fit a wide variety of classroom situations, grades, and age levels. Feel free to use some or all of these materials in any order you wish. Put your own twist on things and customize your own microbe curriculum.

Do I have the capabilities and facilities to teach this material?

Yes, you can do it! Don't let lack of experience, lack of confidence or your fear of failure hold you back. After all, science is built on failure. Dive in and try things. You will be a better teacher for it, and you will be amazed at how much your students learn. The activities in this book can be conducted in practically any classroom setting. Furthermore, the activities can be done quite inexpensively. Microscopes that magnify from 100X to 400X are necessary to do some activities.

How is this book organized?

First comes a section of background information to serve as the basis for instruction. This information can be presented to the students at your discretion, based on your students' previous knowledge of bacteria and viruses and on your teaching style. The Teacher Procedures/Information section provides teachers with an objective for each student activity, background information, and detailed setup instructions for any procedures. The student section outlines the problem to be investigated, gives details necessary for conducting the investigation, tests student understanding with a series of relevant questions, and challenges students to investigate further. Answers for student activities are at the end of each section, followed by full-size graphics that can be posted or used as transparencies. An appendix at the end of the book provides details on commercial suppliers of microbiological materials.

How will my students benefit from this book?

The information sections will provide interesting details and facts, and the activities sections will present students with hands-on, minds-on challenges concerning both viruses and bacteria. Furthermore, the activity sections will develop student process skills in

- **Observing/Describing**
- **Comparing/Contrasting**
- **Experimental Design**
- **Analyzing/Problem Solving**
- **Predicting/Drawing Conclusions**
- **Communicating**

What if I need help?

You will find the technical service department of all the commercial suppliers listed in the Appendix very helpful in solving technical problems or suggesting specific methods. Support and assistance might also be provided by experienced teachers in your system, in your area, or in the microbiology department of local colleges. I love "talking shop" with fellow science educators so if you have a problem, suggestions for improving this book, or neat ideas/activities you think we should incorporate into future editions, please contact me at the address on the title page of this book.

Good Luck and may your trip into the microcosmos be a fruitful one!

SECTION ONE

Plague Watch

1

PROTECTING OURSELVES: IS A PLAGUE ON THE WAY?

Ebola, AIDS, Legionnaire's, E. coli 0157:H7, Hantavirus, Lyme, Bird Flu, and others toll a nightmarish litany of human suffering and death. These new diseases have seemingly appeared from nowhere. In fact, dozens of new and often virulent (deadly) diseases have appeared in the past two decades. According to the National Center for Infectious Diseases, infectious diseases are the third leading cause of death in the U.S. and the leading cause worldwide.

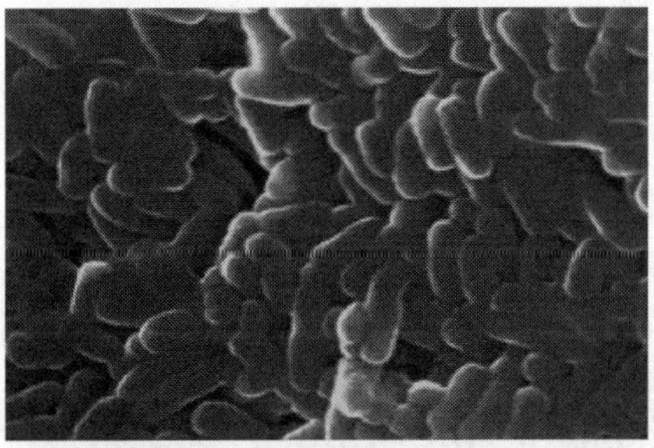
Tuberculosis bacteria

To make matters worse, diseases we once thought were conquered seem to be back with a vengeance. Pneumonia, tuberculosis, meningitis, even children's ear infection—diseases once easily cured with antibiotics—are becoming increasingly resistant to antibiotics. In fact, for some infections we have only one drug left. Once the microbes become resistant to that drug, then we will basically be back where we started.

Another frightening worry is that a superplague may be released as the result of germ warfare experiments by military forces or terrorists.

Where do these new infectious diseases come from, and why have some returned to their former virulence? The answer is starkly simple: a brand new generation of bacteria or viruses can appear in less than an hour. Microbes have accelerated evolution on their side. With this trick, they are well-suited to outwit us and our drugs.

How do microbes enter our body?

There are several portals into us humans:
- the cut, puncture, or insect bite route—tetanus and bubonic plague
- the oral (mouth) route—salmonella (food poisoning) and hepatitis A
- the respiratory route—tuberculosis, flu, and measles
- the genitourinary (sex organs and urinary system openings) route—AIDS and herpes

How do invading parasitic microbes hurt us?

First the microorganism must attach itself to the surface of a cell by binding to specific receptors on the cell surface. Once attached, some bacteria just stay outside the cell and multiply there, often producing toxins (poisons) that kill the cell (as in cholera, strep, or staph infections). Other bacteria—and all viruses—actually invade the cell. Once inside, they multiply, destroying the cell. The new microbes break out to invade other cells.

How do we fight them?

- **Natural Defenses**—Skin and mucous membranes form the body's first line of defense. If these barriers are breached, other nonspecific defenses appear suddenly, including macrophages (white blood cells that swallow up and kill some bacteria on contact). Several days later, if the infection persists, the immune system produces weapons tailored specifically against particular antigens (molecules that the body recognizes as foreign invaders). This arsenal includes B cells—white blood cells that release specific antibodies that bind with particular antigens and destroy them—and T cells—another kind of white blood cell that attacks and destroys any cells that bear antigens they can recognize.

- **Drugs**—Penicillin and some other antibiotics kill bacteria outright; they prevent the bacteria from making new cell walls or interfere with the production of bacterial DNA. Other types of antibiotics do not kill but cripple the bacteria enough for the body's immune system to finish the job. Unfortunately, antibiotics do not work at all against viruses, and many bacteria are becoming immune to the antibiotics we currently employ.

- **Vaccines**—Vaccines prevent disease by stimulating the immune system to produce specific antibodies against particular bacteria or viruses so that the body is always prepared to attack these invaders if they appear. Vaccines consist of either live but harmless microbes, killed microbes, or components of microbes. Unfortunately, the vaccine weapon will not work against many types of disease microbes.

How do the microbes fight back?

- **Against Our Immune System**—Microbes such as HIV, the virus that causes AIDS, constantly vary some of the antigens on their surface to stay one step of ahead of slowly developing (and specifically targeted) antibodies.
- **Against Drugs**—Many microbes rapidly become resistant to drugs through mutations. This enables them to destroy or inactivate antibiotics or to prevent the drugs from reaching key targets inside them. Once they become resistant to a drug, bacteria can share this ability with others through conjugation.

How can the human race survive this microbial onslaught? The Center for Disease Control and Prevention recommends:

- **Surveillance**—Eight specialized programs to identify infectious diseases now exist in the U.S. We need more. If we have global surveillance, we can respond to an outbreak before it spreads.
- **Research**—New vaccines and antibiotics are needed, along with better ways to diagnose and detect disease.
- **Prevention**—Stronger public health systems are needed to provide for adequate restaurant inspections, water-supply tests, and public education. We only spend about 1% of our annual health-care budget on prevention.
- **Education**—The public must play a more active role in the prevention and control of infectious diseases. That means we need better education on the dangers of high-risk behavior and antibiotic abuse as well as the importance of hygiene. Washing hands is still one of the most effective ways to stop illness and disease.

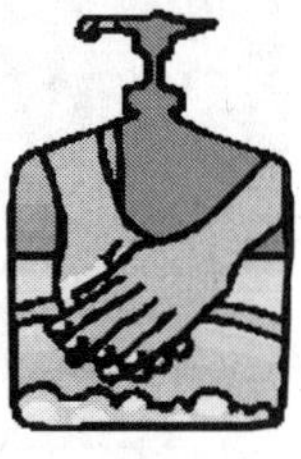

We have to be prepared to deal with the unpredictable. When the next invader appears, we have to be ready to respond.

To prevent infectious diseases ...
(include any or all)

1. **VACCINATE** all children, adults and even pets.
2. **WASH HANDS** often, especially in cold and flu season.
3. **PREPARE FOOD** carefully and cook it thoroughly.
4. **USE ANTIBIOTICS** exactly as directed by your doctor.
5. **REPORT AN INFECTION** that doesn't heal to a doctor.
6. **AVOID AREAS** with lots of insects; use bug repellent.
7. **BE CAUTIOUS** around all wild or unfamiliar animals.
8. **AVOID** unprotected sex and intravenous drug use.
9. **ASK ABOUT DISEASE** threats when you travel.
10. **ALLOW YOURSELF TIME** to recover when sick.

SECTION TWO

Viruses

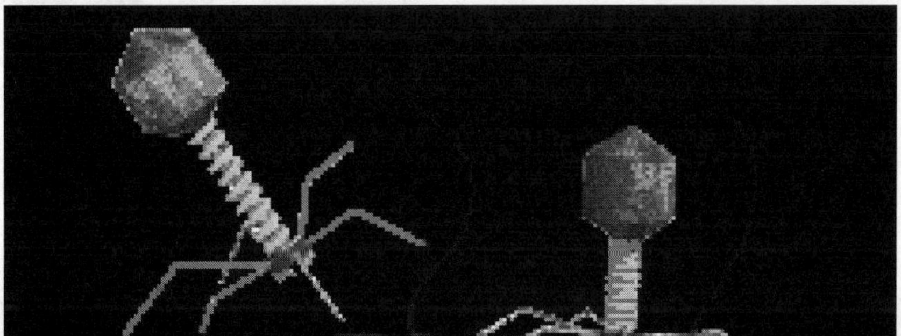

MYSTERY KILLER

Marilyn followed a precise cleanup routine in Johannesburg, South Africa. She was clad from head to toe in protective clothing that made her look more like a futuristic space explorer than the trained medical professional she was.

Why such extreme precaution? The man operated on in that room earlier possibly had viral hepatitis or perhaps worse. Unfortunately for Marilyn, it was worse. Lurking in the dried blood on the instruments, hidden among the folds of soiled, bloody linens, and perhaps even floating in the air itself was the deadly Ebola virus.

The Ebola virus, known as Ebola Hemorrhagic Fever, is one of the deadliest diseases known to humans. It kills 80 to 90 percent of its victims, often in as little as 20 days. This twisted threadlike killer lurks in unknown places, occasionally appearing from nowhere to kill and then mysteriously disappearing again.

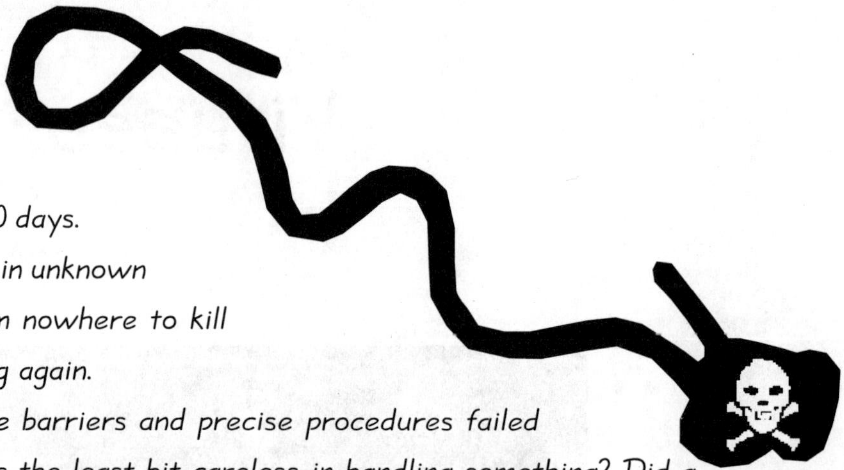

Somehow all the protective barriers and precise procedures failed Marilyn that day. Was she perhaps the least bit careless in handling something? Did a speck of blood find a tiny hole in her clothing? Or did she merely breathe in the virus? No one knows. But once inside her body, the virus quickly went to work.

Within a few hours after exposure, Marilyn began to feel a general weakness. This fatigue quickly progressed into flu-like symptoms—muscle pain, fever, headache, and sore throat. The Ebola virus was now in control, making copies of itself as it destroyed more and more cells. Marilyn was in more pain than she ever imagined possible. Her entire body shut down, and she died less than a month after the invader first slipped into her body.

Ebola is quick, efficient, and devastating. It kills randomly, without warning. And yet, as horrible as it is, the Ebola virus actually kills only hundreds of people at a time (at least so far). The more commonplace measles virus, on the other hand, kills more than 1 million people a year.

2

VIRUSES: PARISITIC PARTICLES

The Organization of Life on Earth

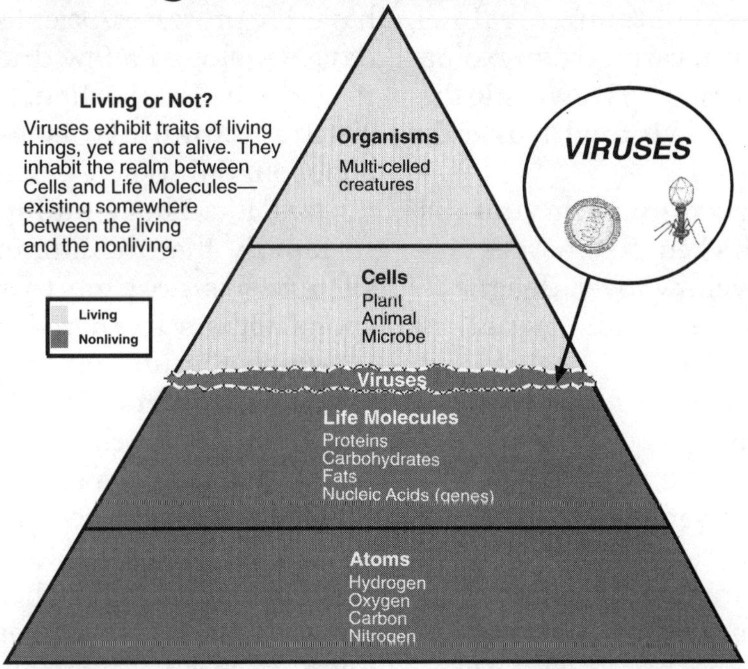

The chart above shows the organization of life on Earth. **(See page 32 for full-size chart.)**

We have identified 92 kinds of atoms (elements), but only four play a major role in life. These atoms are arranged into life molecules—proteins for building structures, carbohydrates for quick energy, fats for stored energy, and nucleic acids for regulation and control.

The molecules of life are, in turn, arranged into living units known as cells. Animal and plant cells contain the same basic parts, but plant cells possess several additional structures that animal cells do not—green "food factories" called chloroplasts and rigid box-like coverings called cell walls. Microbial cells (bacteria) lack many cell parts and will be explained in more detail further on in this book.

Viruses—powerful particles that can have a tremendous impact on living things even though they themselves are not alive—have changed the course of human history numerous times.

The great influenza pandemic of 1918–9 resulted in the death of an estimated 20 million people worldwide—more than were killed during all of World War I.

In our own time, AIDS hysteria has shaken legal, social, and medical tenets to their very core.

Effects of Viruses

Viruses normally have harmful effects on their host cells and are associated with disease. Viral plant diseases of tobacco, cucumber, tomato, lettuce, cabbage, potato, sugar cane—along with potato leaf curl, curly top in beets, and aster yellow—cause great loss of crops and food each year.

Animals also suffer. Many livestock and other economically important animals as well as pets perish each year from such diseases as hog cholera, foot-and-mouth disease, distemper, rabies, swine flu, carp pox, myxomatosis in rabbits, Newcastle disease in fowl, fowl leukemia, and psittacosis in birds.

Humans are plagued by such viral diseases as smallpox, chicken pox, herpes simplex I and II, rabies, yellow fever, dengue fever, rubella, influenza, colds, hepatitis, mumps, hanta, and certain types of cancer.

Disease occurs as viruses damage and kill the cells they invade. However, our bodies are not defenseless against viruses. Viruses must have a weakness or our immune systems could not attack and defeat them. In some cases, our immune system recognizes foreign invaders and manufactures antibodies to attack them. Scientists discovered they could fight certain viral infections by boosting the natural workings of our immune system. Vaccines containing a weakened strain of a virus were created. The vaccine was then injected, stimulating the immune system's manufacturing of antibodies that remain in the blood to attack any further invasions by that specific type of virus. Vaccines have been effective against viruses like polio and smallpox.

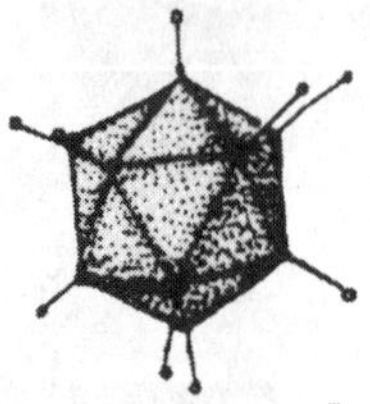

Common Cold Virus

Viral infections must run their course once they occur because no cure has yet been invented. Antibiotics are not effective against viruses because the antibiotics are looking for something viruses don't have: life processes. Scientists have developed a few drugs to slow viral replication, but natural immunity is still about our only defense. Some viruses, like Ebola, multiply so rapidly that the immune system has no time to respond. Others, like herpes, lie dormant in cells to which the immune system does not respond. Still others, like influenza and cold viruses, mutate quickly, making vaccines ineffective in a short time.

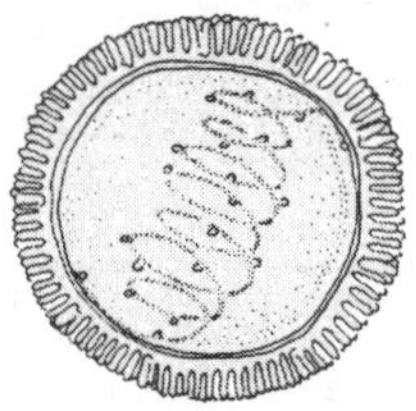

Flu Virus

One positive aspect of viruses is that because it is possible to study the genes that viruses bring into cells when they infect them, viruses have been extremely valuable in genetic research. Many of the major breakthroughs in genetics and molecular biology have been made by studying viral genes. Furthermore, attempts are being made to correct genetic defects by using human-created viruses to insert proper-functioning genes in cells.

(See related student activities p. 17–25)

Antibiotics are not effective against viruses because the antibiotics are looking for something viruses don't have: life processes.

Characteristics of Viruses

As virologists delve deeper into the mysteries of viruses, the following facts have emerged:

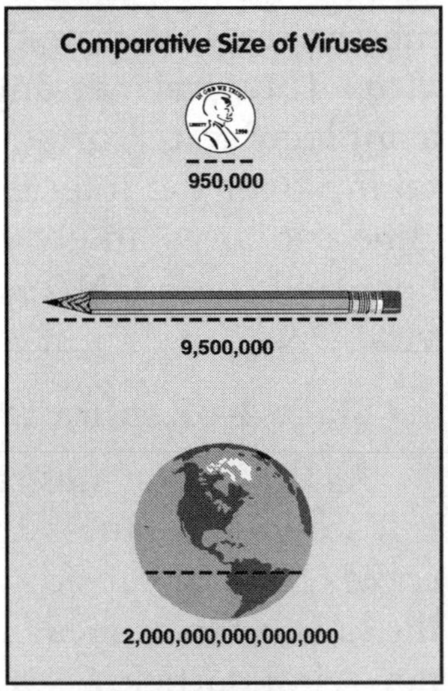

(See page 33 for full-size chart.)

• **Size**—Viruses are so small that special size scales have been developed to measure them. Virologists describe their size in nanometers (1 nanometer [nm] = one billionth of a meter). To help us to visualize such small sizes, let us compare the size of viruses to the sizes of common objects: 950,000 viruses end-to-end would fit across a penny, 9,500,000 viruses end-to-end would fit along the length of a new pencil, and to lay viruses end-to-end around the world at the equator would require about two quadrillion viruses. Ponder this: If all the people in the United States were shrunk to the size of viruses, they would fit on the eraser ends of two pencils with plenty of room left over for future generations.

• **Shape**—Viruses come in a variety of shapes, including helical, polyhedral, and spherical.

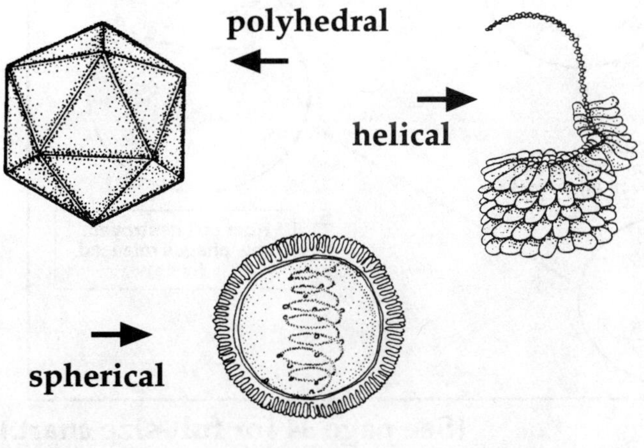

The most complex-looking viruses are the **bacteriophages**, which, as their name suggests, attack bacteria. Resembling little lunar spacecraft, these tiny protein landers have a hollow modular head and spindly legs for grasping the surface of bacteria.

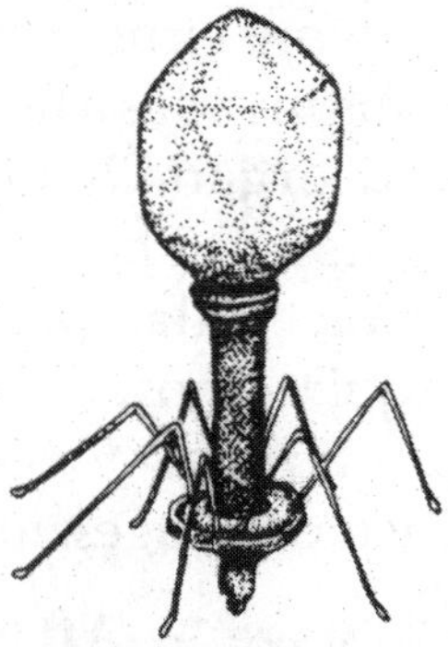

• **Structure**—Whatever their shape, all viruses have the same basic structure. They are nothing more than a protein coat (called a capsid) surrounding a tiny strand(s) of DNA (viral genes) or RNA (chemical codes for making new viruses). The strands of DNA or RNA may be single or double, but the amount of genetic information they carry is small. Most viruses have only about 10 genes; even the largest viruses have only about 100 genes. By comparison, an *E. coli* bacterium has about 2,000 genes, and a human has at least 100,000 genes. Viruses contain no cell parts and are not cellular in nature. Furthermore, they do not require energy. Hence, they take in or make no food and give off no waste products. By these definitions, viruses are not living things.

• **Specificity**—Usually, specific viruses will infect specific organisms. For example, a plant virus cannot infect an animal. There are some viruses that will infect only humans. Others, such as the rabies virus, infect all mammals and some birds. Still others infect only cold-blooded animals. Some infect only closely related species of animals. For example, viruses that infect mice may infect rats. Even the tiniest of creatures, bacteria, many of whom are themselves disease-causers, are infected by bacteriophages (phage from Greek *to eat*).

• **Replication**—Because viruses are not alive, the terms replication and / or duplication are used instead of reproduction to describe the production of new virus particles.

New virus particles are produced when a virus invades a host cell and uses the enzymes and organelles of that cell to make new viruses. They are, therefore, **obligate intracellular parasites,** which means they require a host cell to replicate. The replication of viruses occurs in stages. The T4 bacteriophage will serve to illustrate.

Stage 1—Attachment of host

A virus is activated by chance contact with the right kind of host cell. In the case of T4, molecules on the tail fibers have the proper shape to fit into molecules on the surface of a bacterium. This molecular matching process explains why most viruses recognize and infect only one kind of host cell.

Stage 2—Infection of host

The virus then injects its DNA into the host cell. In most cases, the complete virus particle itself never enters the cell. However, in some cases the whole virus enters, and then the capsid is destroyed.

Stage 3—Sabotage of host's DNA

After entering the host cell, the DNA of the virus goes into action. In most cases, the host cell cannot tell the difference between its own DNA and the DNA of the virus. Consequently, the very same enzyme that makes messenger RNA from the cell's DNA begins to make messenger RNA from the DNA (genes) of the virus. This viral messenger RNA now acts like a molecular terrorist, shutting down and taking over the infected host cell. One viral gene produces an enzyme that destroys the host cell's own DNA but does not harm the viral DNA!

Stage 4—Replication of virus

As the viral genes take over, they use the cell's own structures to make thousands of copies of viral capsids and viral DNA. Soon the host cell becomes filled with hundreds of new viral particles.

Stage 5—Release of virus

The host cell is instructed to make an enzyme that digests the cell wall from the inside. The cell bursts open, releasing hundreds of new viruses. Within three-quarters of an hour as many as 200 new viral particles may burst forth from the microbe. These young phages drift off to infect more bacteria; the unfortunate host, blown to bits, rapidly dies.

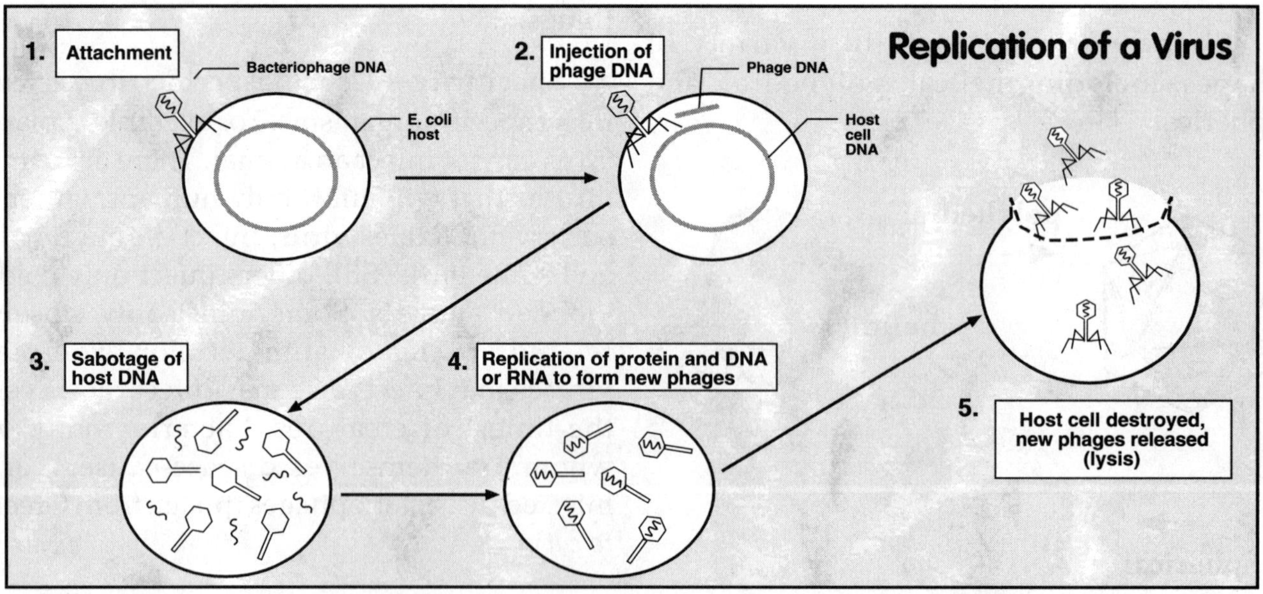

(See page 34 for full-size chart.)

One group of viruses, known as **retroviruses,** do things differently. Retroviruses contain RNA as their genetic material. When retroviruses infect a cell, they produce a DNA copy of their genes. This DNA is inserted into the DNA of the host cell, where it produces the viral messenger RNA that takes over the cell machinery. Retroviruses get their name from the fact that their genetic information is copied backwards—that is, from RNA to DNA rather than from DNA to RNA. Retroviruses cause some types of cancer in humans and animals, and one retrovirus produces AIDS.

A replication pattern like that of T4 bacteriophage is called a **lytic infection** (lysis means to burst open). Some viruses have a modified lytic cycle. Human influenza (flu) virus, for example, does not lyse the host cell. Rather, it exits the host cell by pushing out through the cell membrane of the host. The outer envelope of the virus then consists partly of the host cell membrane and partly of viral protein embedded in the membrane.

Not all viruses cause lysis immediately after entering a cell. Some enter the host cell and become inactive. The viral DNA becomes part of the DNA of the host cell but does not take over the metabolism of the cell right away. When the host cell replicates its DNA, the viral genes are also duplicated. Quietly, the viral DNA may be passed through many generations of the cell with no harm. In this form, the viral DNA is called a **prophage.** Viral inactivity such as this is called a **lysogenic infection.** Eventually, the sleeper awakens and the host cell is destroyed.

Virologists believe some outside influence such as temperature change, certain chemicals, or radiation triggers the prophage.

(See related student activity, p. 17)

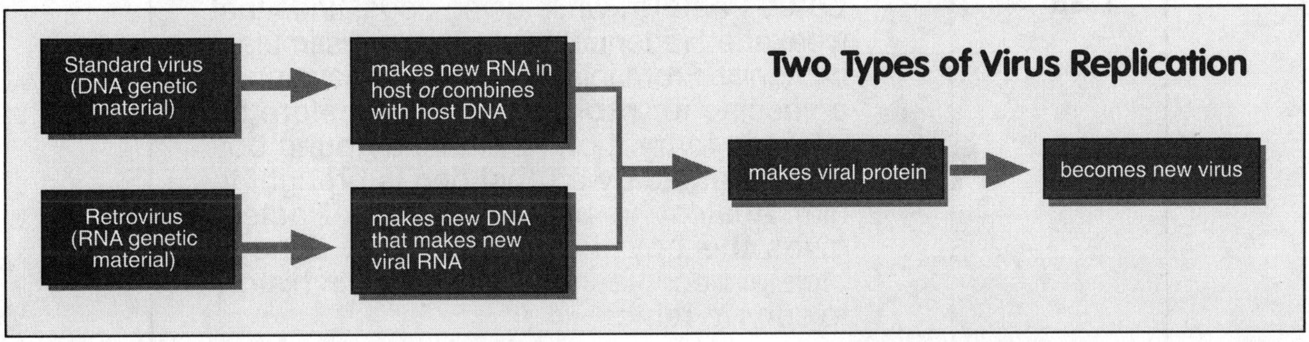

(See page 34 for full-size chart.)

Sources of Viruses

No fossil evidence of viruses has ever been found. Because they are obligate intracellular parasites, viruses probably did not arise until cells appeared, since their existence requires cells. If this is so, then viruses probably either formed spontaneously from existing nonliving organic material or developed as simplification of previously existing cells.

This matter has been further complicated by the discovery of certain disease-causing particles that are even smaller and simpler than viruses. One of these is the **viroid.** A viroid is a short, single strand of RNA with no surrounding capsid. Virologists suggest that somehow this RNA strand interferes with normal cell function and causes production of new viral strands. Diseases caused by viroid infection damage potatoes, coconuts, chrysanthemums, and citrus crops.

Even smaller and simpler than a viroid is a **prion**. A prion doesn't even contain true DNA or RNA but only about 250 amino acids. Yet these tiny molecules have been implicated in **scrapie**, a slow degeneration of the nervous system in sheep and goats; in **kuru**, a degenerative nerve disease contracted by touching the brains of dead individuals; and **mad cow disease**.

(See related student activities, p. 26–7)

Researchers theorized the existence of viruses and even developed vaccines against them long before they were actually seen. The graphic below details some of the history of virus study and vaccines.

Viruses and Vaccines through History

1796 English doctor **Edward Jenner** develops a vaccine against the dreaded scourge of **smallpox** after noticing that milkmaids who contracted cowpox never got smallpox. He correctly surmised that the milkmaids had developed an immunity to smallpox as a result. Taking material from a cowpox sore on the hand of a milkmaid, he inoculated his own son with the first viral vaccine in history.

1885 **Louis Pasteur**, while using dogs, finds that weakened (attenuated) viruses will stimulate immunity. From this work, he has developed a vaccine for **rabies**. However, before he is ready to try it on humans, a young boy badly mauled by a rabid dog is brought to him. Against his better judgment, Pasteur gives the boy the inoculations. Joseph Meister becomes the first person in history to survive rabies.

1892 Russian biologist **Dmitri Iwanowski** discovers the virus that causes **tobacco mosaic disease** in plants.

1898 Dutch biologist **Martinus Beijernick** coins the term virus (from Latin for *poison*) and shows viruses to be infectious "germs" by transferring viral diseases from sick to healthy plants.

1930s The first electron microscope is perfected.

1940s The first virus is actually seen through the newly developed electron microscope.

Teacher Procedures/Information

Medical Situations

Objective: Students will point out differences between viruses and bacteria, in part by describing treatment for viral infections.

Student Activities are on pages 17–8.

Answers appear on page 29.

Viruses Concept Map

Objective: Students will complete a visual map to demonstrate understanding of key concepts and interrelationships of viruses.

Student Activity is on page 19.

Answers appear on page 30.

Disease Detectives

Objective: Students will exchange test tube fluids to simulate an epidemic of infection; they will identify the "carrier" by tracing the path of infection.

Student Activities are on pages 20–1.

Answers will vary according to student experience.

A. Background

From bubonic plague to AIDS to influenza, scientists and public health officials have struggled to understand and contain the spread of infectious diseases. Using common laboratory equipment and solutions, students can simulate the spread of a simple imaginary disease in order to explore some of the factors that affect the rate of infection and the difficulties in tracing the path of transmission.

B. Materials

- a numbered test tube and a dropper for each student
- 0.1 molar NaOH solution
- phenolphthalein solution
- distilled water

C. Procedure

1. Prepare the test tubes before class: Fill one tube halfway with 0.1 molar NaOH solution; fill the rest of the tubes halfway with distilled water only. You can prepare a 0.1 molar NaOH solution by dissolving 4.0 grams of NaOH (sodium hydroxide) in 1 liter of distilled water. If you do not have access to the chemicals needed for this activity or if you have no experience mixing solutions, perhaps the chemistry department at your school or area high school could help. The necessary chemicals can also be purchased from the suppliers listed in the appendix section of this book.
2. On the day of the activity, distribute test tubes and droppers randomly to the class. Make a mental note of which student receives the test tube containing the NaOH solution. The students should not be able to tell because both the distilled water and the NaOH solution will be clear. ***Caution***—Sodium hydroxide (NaOH) and phenolphthalein can irritate the eyes and skin. Alert students to avoid spilling and contacting the solution in their test tube.
3. Have students walk around the room with their test tubes and droppers. When you say "Stop," each student should use his or her dropper to trade a drop of fluid with the nearest person. Students should record the number of each student they trade with (contact). Repeat this until three trades have occurred. If you have a small class, say 15 or less, you might consider having the students doing only two trades. Otherwise, you run the risk of *all* the students becoming infected very quickly.
4. Now it is time to test for the imaginary infection. Put a drop of phenolphthalein in each test tube. Phenolphthalein is a clear indicator liquid that turns pinkish purple in

Teacher Procedures/Information

Disease Detectives (cont.)

the presence of NaOH. Therefore, if the fluid in a student's test tube turns pinkish purple, that student is "infected."

D. Examination of Results

Challenge the class to now become epidemiologists and determine who the original carrier of the "disease" was. Point out to students that the simulated disease has a 100% rate of infection that appears immediately upon testing. Therefore, each person who received the infected fluid can be identified immediately. By then checking each student's contact record, you can determine which student spread the disease. Some diseases, such as AIDS and chicken pox, can remain dormant in the body for a long time. Others, such as Ebola, kill the host rapidly. Ask students to consider how each of these factors affects the spread of disease and our ability to identify carriers.

E. Additional Student Activities

1. Investigate why a person who is the original carrier and transmitter of a disease is sometimes called "Typhoid Mary." Write a short report on this topic.
2. Discuss within your group how the class acted toward the "carrier" once you determined who he or she was. Was there a social impact on the "carrier"?

F. A True Story

Mary Mallon was an immigrant Irish cook. Between 1900 and 1907, she infected twenty-two New Yorkers with typhoid fever through her puddings and cakes; one of them died, but Mary never showed any symptoms of the disease. Tracked down through epidemiological detective work, she was apprehended as she hid behind a barricade of trash cans. To protect the public's health, authorities isolated her on Manhattan's North Border Island, where she died some thirty years later.

Deadly Mosaic

Objective: Students will form hypotheses and design experiments using plant leaves and a viral agent to form conclusions about viral infection.

Student Activities are on pages 22–5.

Answers will vary according to student experience.

A. Background

Tobacco mosaic is the best-known of a number of viral diseases that infect plants. The virus derives its name from the appearance of the destruction it causes. (The leaves of plants that have this disease become spotted in patterns that look like the designs called mosaics.) As with most plant viral diseases, tobacco mosaic is introduced into plants by insects or through breaks in the plant's tissues, such as a tear in a leaf. Mosaic diseases destroy large amounts of valuable crop plants every year. Tomatoes, peppers, tobacco, beets, watermelons, cucumbers, pumpkins, squash, and lettuce are some of the plants susceptible to tobacco mosaic attack.

B. Procedure

Have students brainstorm the problem they have been given in either small groups or a teacher-directed discussion with the entire class. Whatever the arrangement, the teacher should guide and lead the students to a well designed experiment, not just tell them how to do it.

Teacher Procedures/Information

Deadly Mosaic (cont.)

C. Think about It

1. Start the brainstorming session by having students specifically state what the problem is they are trying to solve. One way of stating the problem at hand might be: **Is mosaic disease caused by an infectious particle?** Make sure students follow the hints for stating scientific problems given in the student section (page 22).
2. Once a scientist has clarified the problem to be solved, the next step is to do a literature search to gather information that might help in designing an experiment to solve the problem. This is an important step, but I suggest you have students forgo it because it is quite time-consuming, and students will find the answer to the problem and not have to solve it for themselves.
3. Students should now state their personal hypothesis regarding the problem. There are only two choices:
   a. yes, mosaic disease is caused by an infectious particle *or*
   b. no, mosaic disease is not caused by an infectious particle.
4. Now comes the difficult part—designing a sound experiment to test our hypothesis. One possible experimental design follows (Teacher demonstrates):
   a. Place pinches of tobacco from different brands of cigarettes into a mortar. Add 5 ml. of potassium phosphate solution and grind the mixture with a pestle.
   b. Pour the mixture into a beaker. This mixture should contain tobacco mosaic viruses and can be used to infect plants. Ask students to consider: **Aside from health and safety reasons, why do greenhouse operators not allow smoking in their greenhouses? Answer:** Cigarette tobacco contains tobacco mosaic viruses and smoking could spread these viruses to the plants in the greenhouse.
   c. You will need at least 3 plants of a type that is susceptible to tobacco mosaic. Using very fine grit sandpaper, lightly rub the upper surface of several leaves of each plant. Ask students to consider: **Why sandpaper the leaves? Answer:** The sandpaper abrades the leaves and facilitates infection.
   d. Using cotton swabs, rub the following into the abraded leaf areas:
      - Plant 1—some of the cigarette tobacco potassium phosphate mixture
      - Plant 2—some potassium phosphate solution only
      - Plant 3—water only

   Ask students to consider: **Which plant(s) is the experimental and which plant(s) is the control?** (Plant 1 is the experimental—what we are testing—while Plants 2 and 3 are the control—for comparison)
5. Place the plants in a suitable well-lit location and keep them well watered. Students should periodically observe the plants and collect pertinent data such as presence or absence of mosaic spots, size of mosaic spots, number of mosaic spots per leaf, and differences in size and number of mosaic spots between experimental plant(s) and control plant(s).
6. Once a reasonable amount of time has passed, students should use their observations and data to draw a conclusion. In other words, what is the answer to the problem you were trying to solve?

D. Materials

If you follow the above experimental design, you will need the following materials:

Teacher Procedures/Information

Deadly Mosaic (cont.)

- at least 3 plants of a type that is susceptible to tobacco mosaic virus (see the Background material for a list of suitable plants). Plants can be started well ahead of time from seed or if in season, purchased from a greenhouse or garden supply firm.
- tobacco from several brands of cigarettes
- 10 ml. of 0.1 M dibasic potassium phosphate solution—Potassium phosphate may be purchased from the suppliers listed in the appendix section of this book or perhaps your chemistry department would have some on hand. Prepare a 0.1 molar solution by mixing 2.1 grams of potassium phosphate in one liter of water.
- mortar and pestle—Again, perhaps your chemistry department would have these.
- very fine grit sandpaper
- cotton swabs

E. The Next Step

If you wish students to investigate further, consider having them design experiments to test which of several types of plants are the most susceptible to the virus or whether age of the plant plays a role. Are young plants more or less susceptible to the virus than older or mature plants?

Computer Viruses

Objective: Students will write a definition and complete a graphic organizer to demonstrate distinctions and similarities between a computer virus and a biological virus.

Student Activities are on page 26.

Answers appear on page 31.

Putting It on Paper

Objective: Students will use writing (creative or report) to demonstrate knowledge in one of four possible areas:

A. Students will describe the effects of viruses on the human body.
B. Students will extrapolate from real knowledge of human/viral relationships to create a fictional account of a virus from another planet.
C. Students will describe a correlation between human actions and emergence of disease.
D. Students will describe the importance of the human role in the balance of the ecosystem.

Student Activities are on pages 27–8.

Answers will vary according to student experience.

Medical Situations

Rx

1. **Your doctor's diagnosis is that you have a very bad cold, but your doctor does not prescribe antibiotics. Why not?**

2. **So if the doctor won't prescribe antibiotics when you have a cold or the flu, how do doctors treat common viral infections?**

3. **After your trip to the doctor, you are confused about the differences between viruses and bacteria. You do some research and, in the process, you also find information about viroids and prions. Fill in the chart below.**

Virus	Bacteria
Viroid	**Prion**

Medical Situations

Continued

℞

4 **Your aunt Mabel tells you that as a young girl, she had a serious case of chicken pox and that years later, she had a case of shingles. Shingles is a disease caused by the same virus that causes chicken pox. Is it possible for shingles to appear years after a case of the chicken pox? If so, how?**

5 **You and most other people have had childhood vaccinations against serious viral diseases. How do vaccines work?**

6 **You are the famous virologist I.C. Clearly and you wish to replicate some viruses. How would you go about it?**

Viruses Concept Map

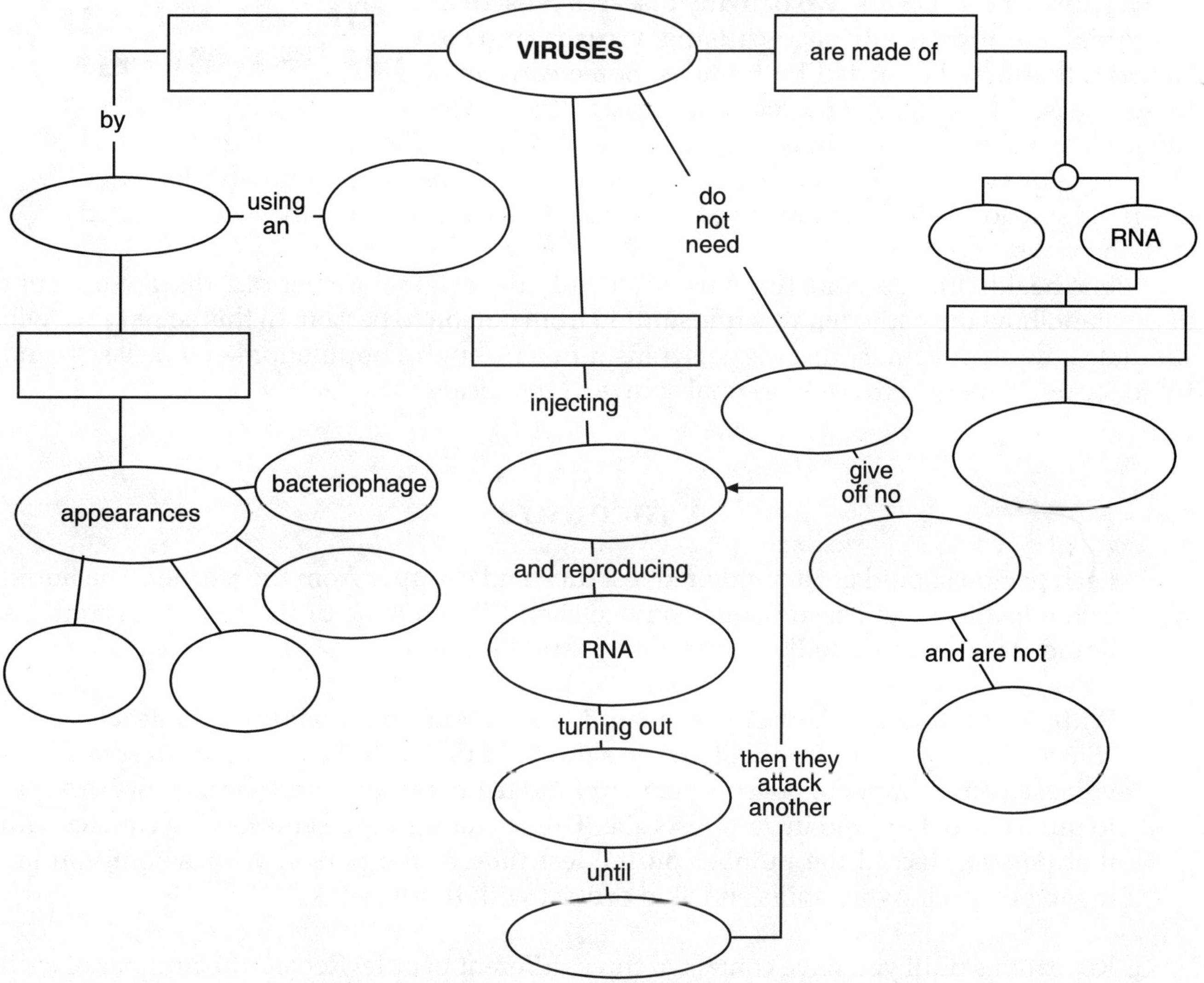

Connectors

can be seen
can identify several
replicate by
surrounded by

Note: ○ represents "or"

Word List

cellular
DNA
capsid
electron microscope
food
helix
host cell
new viral particles
polyhedral
spherical
the cell bursts
virologists
wastes

Disease Detectives

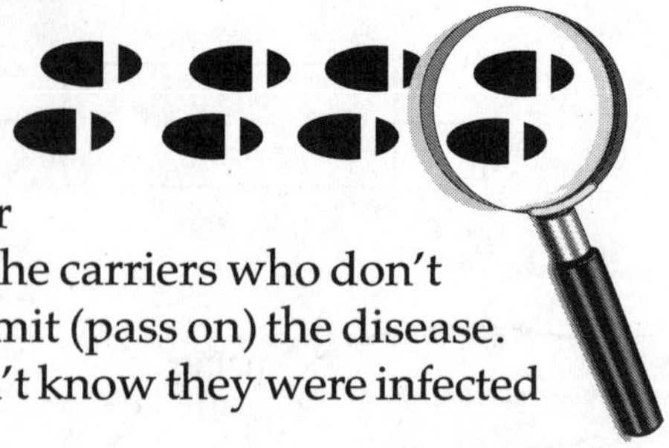

Hepatitis and AIDS are two of many diseases caused by microbial pathogens (disease-causing viruses and/or bacteria) that may be carried by humans. Some carriers of these diseases show the symptoms of the disease (they appear sick), while others show no signs and look perfectly healthy. The carriers who don't show any signs of the disease are the ones most likely to transmit (pass on) the disease. The rapid spread of AIDS occurred in part because people didn't know they were infected with the virus.

Public health officials sometimes try to identify the original carrier of a disease to better understand how the pathogen was transmitted from person to person. In this activity we will simulate how an imaginary disease gets transmitted through a population—your class— and try to locate "patient zero," the original source of the disease.

Procedure

1. Each person should get a numbered test tube and dropper from the teacher. The liquid in the test tube represents your body cells and fluids. One of the test tubes contains liquids that are "infected" with an imaginary disease.

2. With your test tube and dropper in hand, walk about the room until the teacher says "Stop," then use your dropper to trade a drop of fluid with the person nearest to you—you use your dropper to put some of your fluid in her test tube and she uses her dropper to put some of her fluid in your test tube. This exchange represents having contact with that person. Record the number on the test tube of the person with whom you are having contact. Avoid spills and skin contact with fluids.

3. Repeat this until you have contacted three different people. Record the number of each person you have contact with.

4. The teacher will now come around and test each student's fluids for the presence of disease. If your fluids turn pinkish purple, you have the disease; if your fluids remain clear, you do not have the disease.

Disease Detectives

Continued

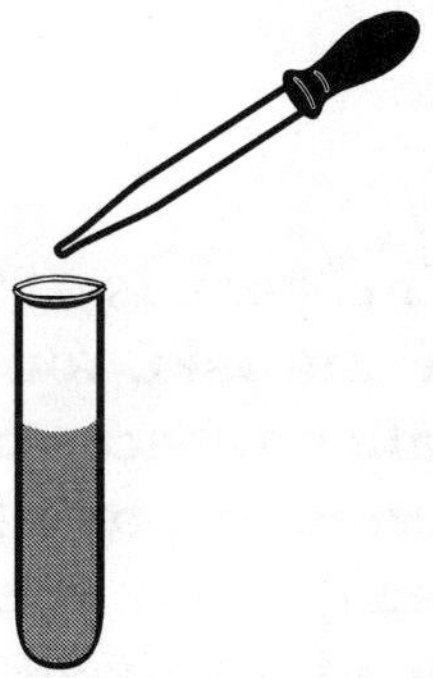

1 How many students are now "infected"?

2 Who started this "epidemic"? Now play the role of epidemiologists (disease detectives). Collect data that will help you trace the path of the epidemic and locate the original carrier—"patient zero." Explain how you arrived at your conclusions.

3 If you did one more round of contacts, how many students would be "infected"? How many rounds would it take to guarantee that every person in the class would have the "disease"?

4 Describe some ways that you could have kept from getting "infected" during this activity.

Deadly Mosaic

1 The year is 1892 and you are Russian biologist Dmitri Iwanowski. You have been studying a disease in tobacco plants called tobacco mosaic. At first you thought the disease was caused by bacteria. You tested this by filtering the juice of an infected plant using minute porcelain filters to remove any bacteria and rubbing the filtered juice on a healthy plant. The healthy plant did develop the disease, so you have ruled out bacteria as the cause of tobacco mosaic. Even though you cannot see the cause, your belief (hypothesis) now is that this plant disease is caused by a tiny particle that can carry this infection from plant to plant.

How can you test this hypothesis?

In science (or in any other situation for that matter), you cannot solve the problem if you don't understand it. Here are some hints on stating problems:

UNDERSTANDING THE PROBLEM

1. State the problem in the form of a question.
2. Keep the problem as simple as possible.
3. Make sure you are solving only one problem at a time.

2 Once a scientist has successfully stated and understood his/her problem, the next step is to research what has been done previously regarding that particular problem. Even though such a search is important, why did your teacher recommend that you eliminate this step?

Deadly Mosaic

Continued

3 **Now specifically state your hypothesis about the problem.**

4 **In the space that follows, explain how your experiment will proceed. List all the materials you will use and explain step-by-step how you will proceed.**

Materials Needed

Step by Step

1.

2.

3.

4.

5.

6.

Deadly Mosaic

Continued

5 Now set up and conduct your experiment. As your experiment progresses, use the space below to keep observations and important data.

Deadly Mosaic

Continued

6 **Once a scientist has collected enough data or all the data possible, he/she forms a conclusion. What is your conclusion? Or, what is the answer to the problem you were trying to solve?**

7 **Defend your conclusion. What observations and data led you to form your conclusion?**

8 **Tobacco mosaic virus is capable of infecting different species of plants. Design an experiment to determine which of several types of plants are most susceptible to the virus.**

Computer Viruses

1 **What is a computer virus?**

2 **Relate the action of a computer virus to that of a biological virus.**

COMPARE AND CONTRAST

Biological virus | **Computer virus**

HOW ALIKE?

HOW DIFFERENT?

WITH REGARD TO

CONCLUSION OR INTERPRETATION:

Putting it on Paper

A. A Formal Library Report

Pick a type of virus from the following list and write a report on that virus.

adenoviruses	influenza	smallpox
rabies	HIV	polio
rubella	hepatitis	mumps
herpes	Ebola	chicken pox

The report should include these things:

1. which organ systems of the body are attacked and what happens to them
2. the physical symptoms
3. the way(s) the disease is spread
4. whether a vaccine is available against the disease
5. how the symptoms of the disease are treated
6. the usual outcome of having the disease

B. A Virus Story

In *The War of the Worlds*, H.G. Wells told the story of an invasion of Earth by aliens from Mars. Our weapons proved useless against the Martians, who devastated the planet. But in their moment of triumph, they were cut down by simple bacterial diseases, against which they had no immunity.

Using a similar theme, write a story about humans voyaging to another planet sometime in the future. In this story, have the humans contract an alien virus. How the humans get the virus and what it does to them is up to you and your imagination.

Putting it on Paper

Continued

C. An Essay on Mad Cow Disease

Mad cow disease, a brain-wasting illness now linked to prions found mostly in British livestock, emerged as part of the agricultural industry's response to certain economic pressures. For many years, cattle food was mixed with protein derived from sheep. The protein was obtained as a by-product of another process used to obtain fat from sheep. This fat was used to manufacture wax.

When prices for wax were high, expensive solvents were used to process the sheep carcasses. This solvent apparently destroyed a naturally occurring prion that had co-existed with the sheep without causing disease. After wax prices fell, expensive solvents were no longer used. As a consequence, the prion was not destroyed and made its way into the cattle feed, infecting thousands of cows with a disease that attacks brain cells.

This is an example of the transfer of a prion that once co-existed with one species but then leaped to another species with disastrous results.

Write an essay on this situation in which you consider the following questions:

1. What does the emergence of mad cow disease illustrate about the consequences of human actions?
2. Can you find other examples in which the economic or social demands of human needs have changed the balance in an ecosystem, resulting in the emergence of disease?

D. What's in a Quote?

React in writing to this quote:

> **"The single biggest threat to man's continued dominance on the planet is the virus."**
>
> — Joshua Lederberg, bacteriologist and Nobel laureate

Answers

Medical Situations

1. **You have a very bad cold, but your doctor does not give you antibiotics. Why not?**
Antibiotics are chemicals, derived mainly from molds, that kill bacteria in people and animals. Since a virus causes a cold and viruses are not alive, a chemical designed to kill living cells is useless against them.

2. **So if the doctor won't prescribe antibiotics when you have a cold or the flu, how do doctors treat common viral infections?**
Common viral illnesses such as colds, influenza, and chicken pox can only be treated with rest, fluids, vitamins, medications that relieve the symptoms, and time—time for the body's immune system to fight off the virus.

3. **After your trip to the doctor, you are confused about the differences between viruses and bacteria. You do some research and in the process you also find information about viroids and prions. Fill in the chart below.**

Virus—A virus is nothing more than an extremely tiny particle of protein covering a few genes (RNA or DNA). Viruses take in no food and give off no waste products. They duplicate by taking over living cells and forcing those cells to make more viruses. By any of these definitions, viruses are not alive.

Viroids—Viroids are even simpler than viruses, being nothing more than a short, single strand of genes (RNA) with no covering protein capsule.

Bacteria—Bacteria are living creatures. Their whole body is nothing more than one very small cell; but with that one small cell, they carry out all the life processes of larger organisms—they take in or make food, they give off waste products, they grow, and they eventually die.

Prions—Simplest of all, prions contain no genes or protein capsule. They are merely a strand of about 250 amino acids.

4. **Your aunt Mabel tells you that as a young girl, she had a serious case of chicken pox and that years later, she had a case of shingles. Shingles is a disease caused by the same herpes virus that causes chicken pox. Is it possible for shingles to appear years after a case of the chicken pox?**
Shingles is caused by the reactivation of the virus that produced the chicken pox in the first place. The virus waits dormant for years in body cells, then is triggered to enter another disease-causing lytic cycle—this time with different symptoms.

5. **You and most other people have had childhood vaccinations against serious viral diseases. How do vaccines work?**
Vaccines contain a weakened strain of certain viruses. The vaccine is injected into the body and, in response, the immune system is stimulated to manufacture antibodies that remain in the blood to attack any further invasion by that particular type of virus.

6. **You are the famous virologist Dr. I. C. Clearly, and you wish to grow some viruses. How would you go about it?**
Viruses cannot be replicated in the laboratory on synthetic media. Instead, they must be replicated on the cells they use to replicate themselves. This involves replicating them in live plants or in live animals such as mice or in chicken eggs.

Answers

Viruses Concept Map

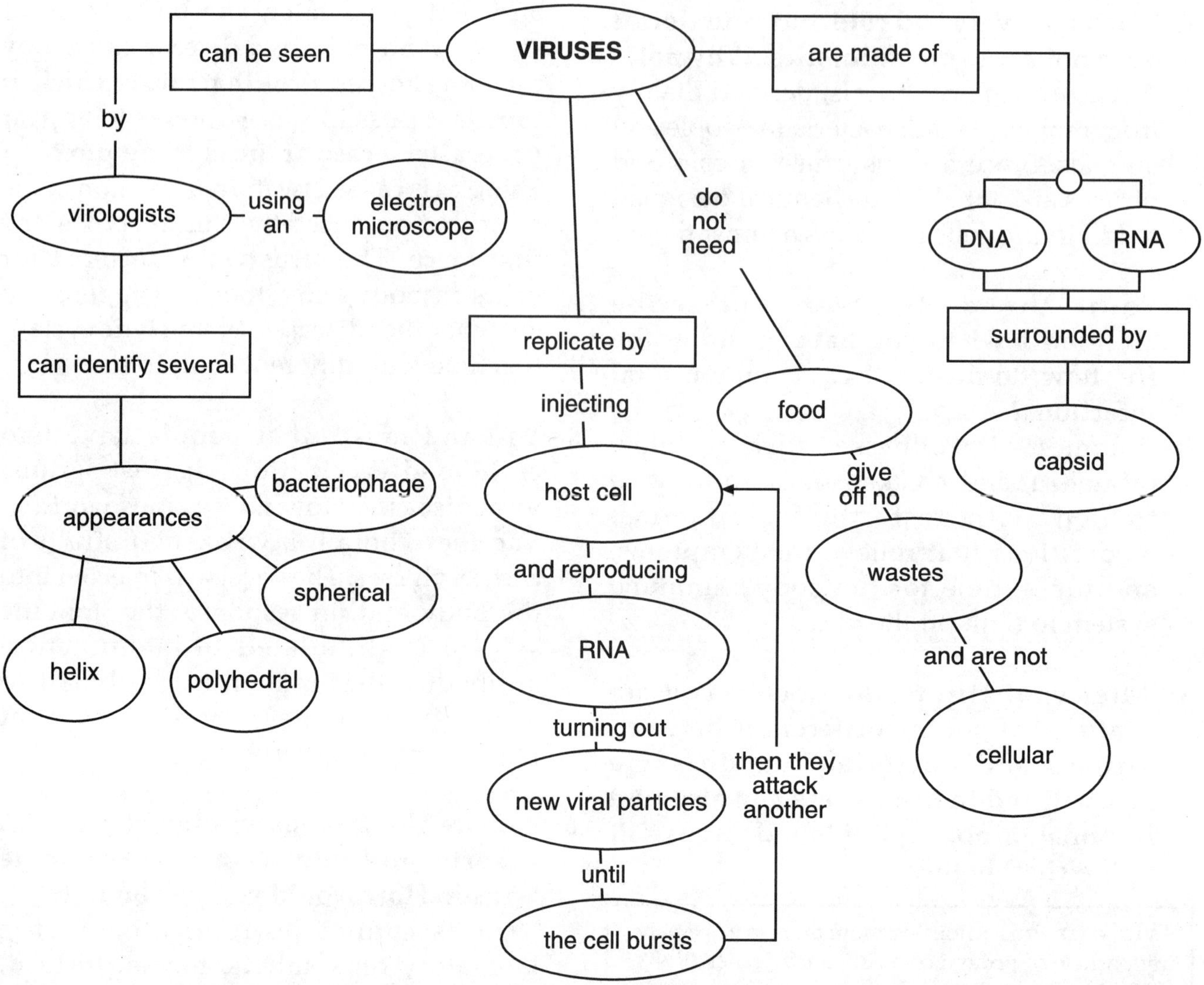

Disease Detectives

Answers will vary according to student experience. Students should be able to support their conclusions with careful observation and collection of data. The only sure way to avoid getting "infected" is to avoid contact.

Deadly Mosaic

Answers will vary according to student experience. Students should be able to support their conclusions with careful observation and collection of data.

Answers

Computer Viruses

1. **What is a computer virus?**
Computers of all sizes from the large mainframes of the government to individual personal computers are increasingly under attack from computer "viruses." A computer virus is a sequence of instructions, secretively inserted in a software program, that causes the computer to halt its programmed activity and replicate the virus's instructions. This usually results in the original program's being damaged or destroyed.

2.

COMPARE AND CONTRAST

Biological virus | **Computer virus**

HOW ALIKE?

- Both are tiny.
- Both insert their program or genes into a larger entity.
- Both are "saboteurs," destroying either programs or cells.
- Both are difficult to combat and cure.
- Both are relatively simple but have the potential to do great harm.

HOW DIFFERENT?

Biological virus	WITH REGARD TO	Computer virus
atoms and molecules	structures	a number code
naturally occurring	origin	human-made
cells	target	larger program
cell destruction	results	program destruction

CONCLUSION OR INTERPRETATION:

Biological viruses are organic and attack living cells; computer viruses are inorganic number codes that attack larger programs.

The Organization of Life on Earth

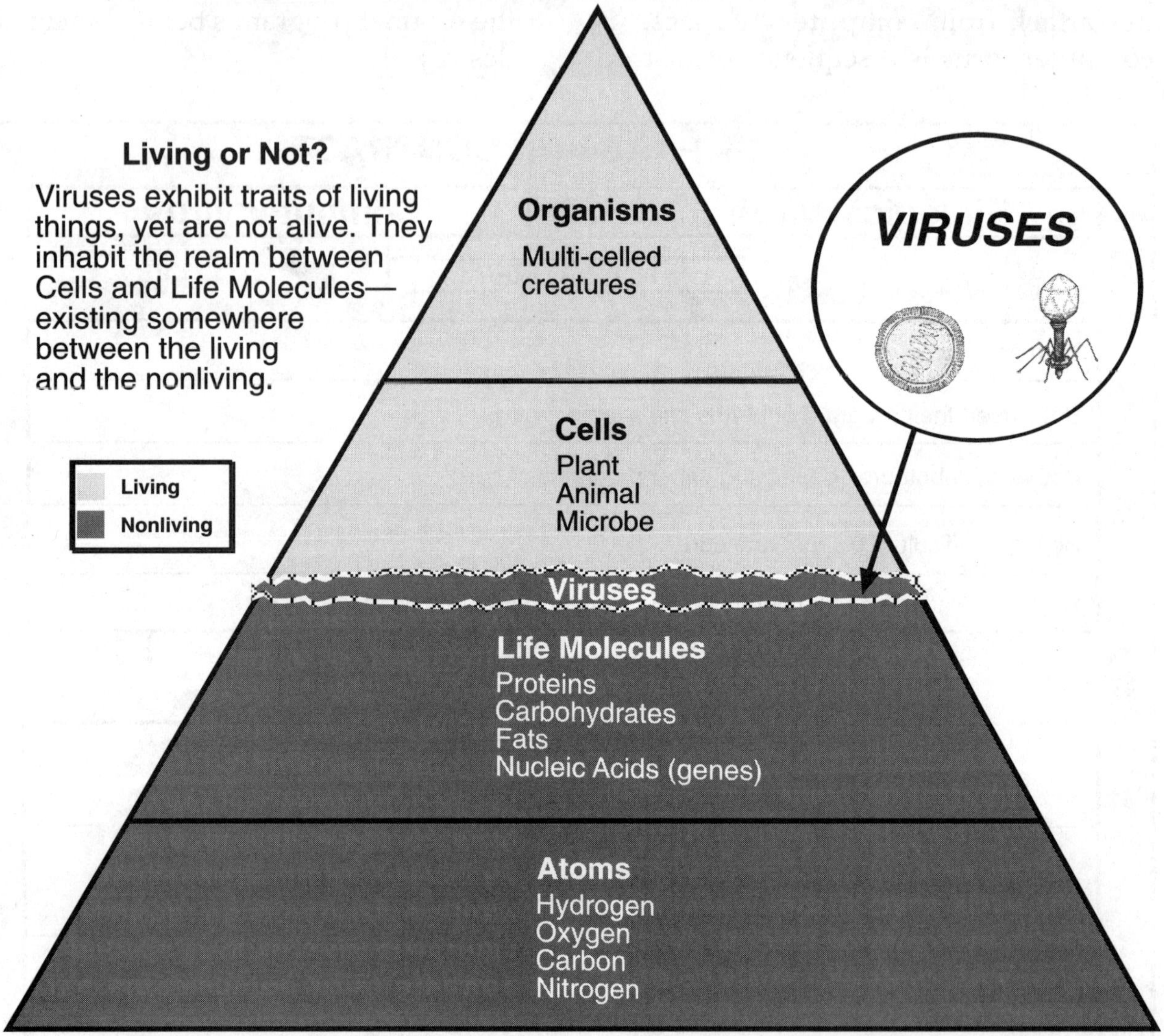

Comparative Size of Viruses

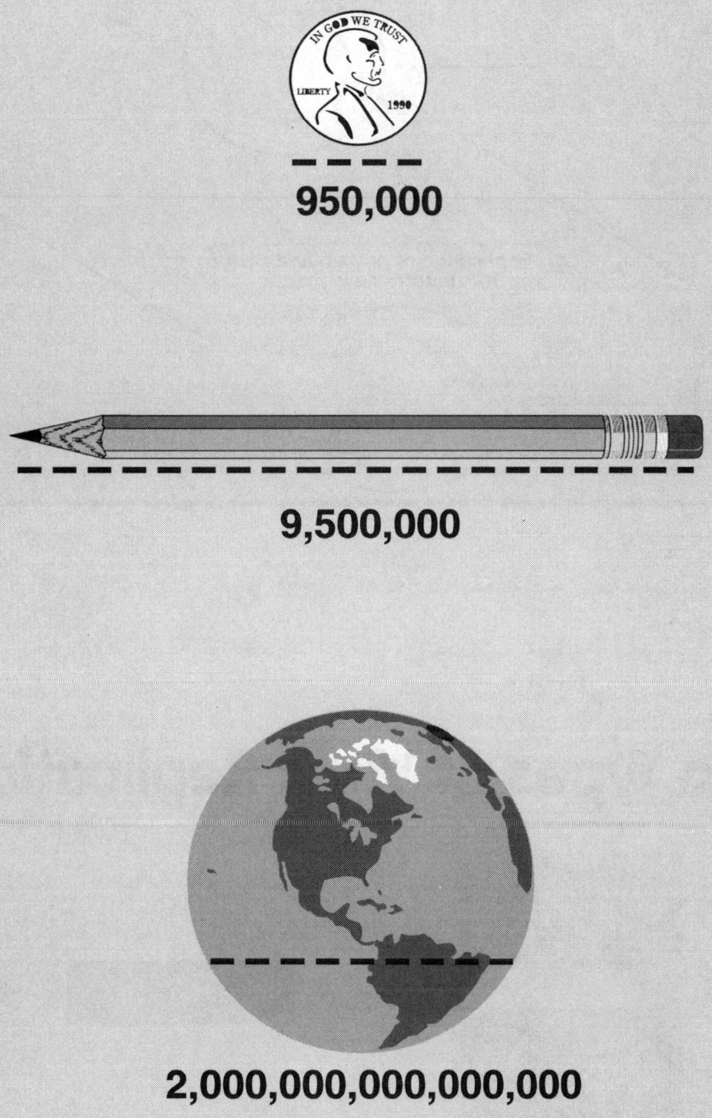

Replication of a Virus

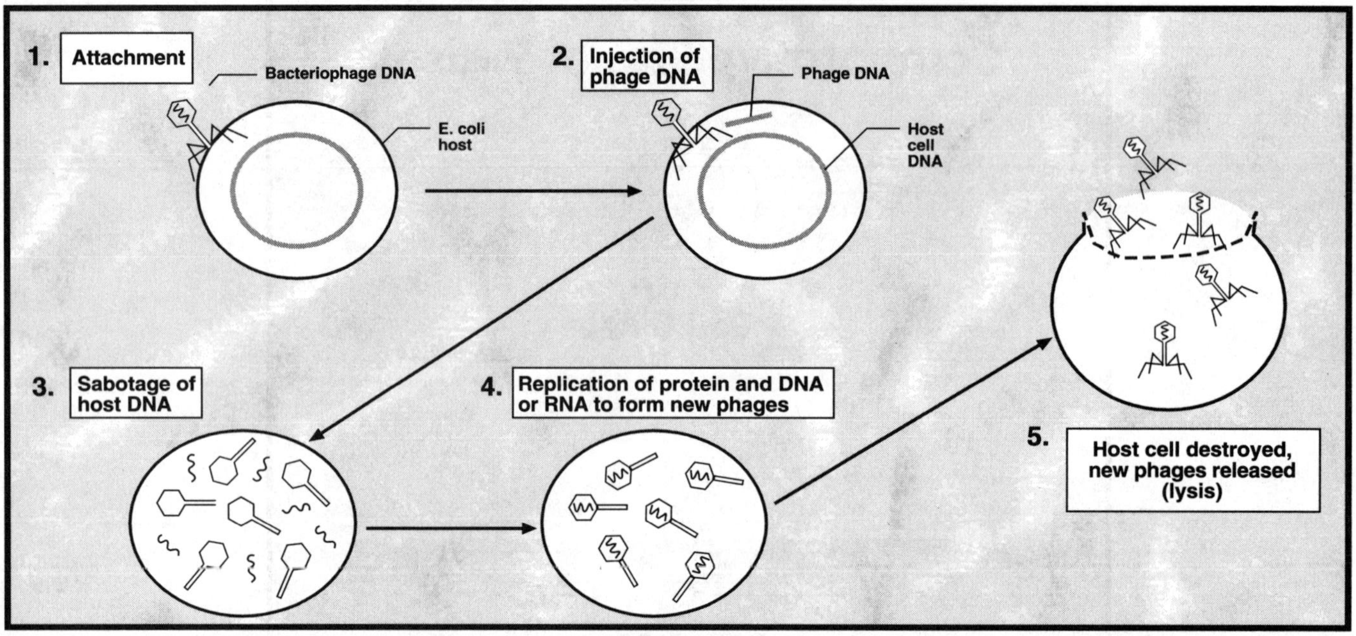

Two Types of Virus Replication

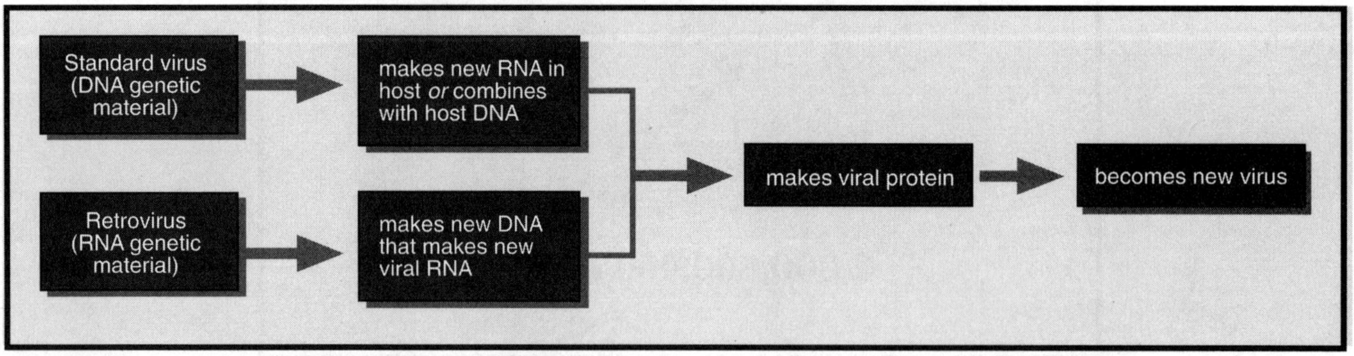

SECTION THREE

Bacteria

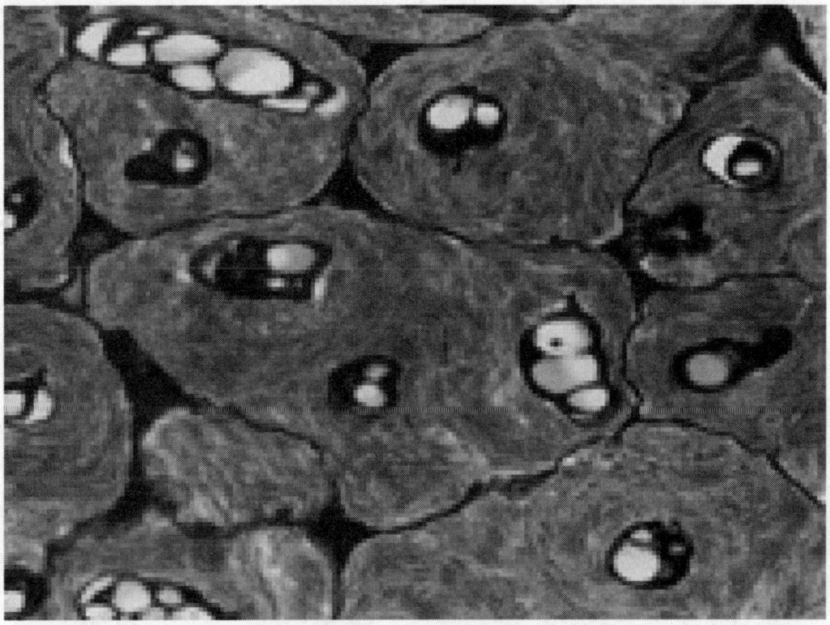

BRUSH WITH DEATH

The pictures of Donna's right hand document her survival against a hideous foe. The before picture highlights a damaged hand with only the little finger intact. The after picture features a normal-looking hand with three prosthetic (artificial) fingers that are so natural in appearance that, "even when I went ring shopping the jewelry salespeople didn't notice and they were staring right at my hands!"

Who could have imagined that someone like Donna, who was the picture of good health, always preaching exercise and proper eating, would have had her life changed nearly overnight by the simplest of creatures—bacteria?

Her ordeal began on a ski trip. Donna noticed a small cut on one finger. She remembers thinking, "I need to clean this cut and get a bandage on it because fuzz from my glove is getting into it."

The next day, her right arm was uncomfortably sore. She thought this strange because her left arm felt fine. By the next evening, she was bedridden, wracked with a cycle of vomiting and diarrhea. Donna and her friends believed she had only a bad case of flu. Little did they know that flesh-eating bacteria had entered the cut on her finger and were beginning to kill the soft tissues of her hand and arm, a condition known as necrotizing fasciitis (meaning decaying skin).

Necrotizing fasciitis (or necrotizing myositis if muscle tissue is destroyed) is caused primarily by Strep A bacteria (the same bacterium that causes common Strep throat), but other types of bacteria or even mixtures of bacteria can also cause it. These bacteria can enter through any opening in the skin regardless of how minor—a paper cut, a staple puncture, or a pin prick—but also through a bruise, blister, or abrasion.

What turns these common bacteria into virulent flesh-eaters? No one knows, but once they get a hold on the body they can quickly become life threatening as they kill soft tissue and cause toxic shock syndrome, both deadly alone but especially devastating together.

Donna flew home so sick she had to be carried on and off the plane. She quickly ended up in the hospital's intensive care ward, being administered a multitude of fluids, antibiotics, and blood pressure medications. (She had no measurable blood pressure when admitted.)

About 3 A.M. on January 31, her vital signs crashed and she was placed on life support. Her family was called in and told she might not pull through. Donna would not regain consciousness again until February 10.

During her long "nap," surgeons operated five times: once to remove dead tissue from her arm, once to cut off the finger where the infection started, once to scrape dead material from several fingers, once to close up drainage areas in her arm, and finally to graft skin from her thighs to her arm.

After months of therapy, Donna slowly regained the strength (no muscle tissue was removed) in her arm, and she received her prosthetic fingers about seven months after her brush with death. Today, she leads an active, normal life and has become a leader in educating people on the nature and prevention of this infection.

As this story illustrates, bacteria can be deadly. They can kill and disfigure many people worldwide each day. However, life itself would not be possible without ordinary decay bacteria recycling dead material back into life-giving nutrients.

3

BACTERIA: TINIEST LIFE

Chapter 1: Bacterial Classification—What Is A Moneran?

People have been classifying (grouping) living things for various reasons since ancient times. One of the first attempts to group life scientifically can be traced to the Greek philosopher/scientist Aristotle, who separated all living things into two kingdoms—plant and animal.

This plant-animal dichotomy served biologists well for centuries. However, in the mid-1600s, a Dutch drapery shop owner named Anthony van Leeuwenhoek, attempting to detect imperfections in fabric, refined a single-lens microscope into an instrument capable of magnifying over 200 hundred times and inadvertently stumbled into an invisible world.

Leeuwenhoek was amazed to see "wee beasties" cavorting in pond water and amazed to find "animalcules" in the tartar from his own teeth. Today we know that Leeuwenhoek was probably the first person to observe and describe living bacteria.

Advances in microscopy swiftly followed and it quickly became obvious that the two-kingdom system was grossly inadequate. Clearly, microscopic creatures differed from true plants and animals and, even in the microscopic realm, bacteria were fundamentally different from all other life forms on Earth.

In 1959, Cornell University biologist R.H. Whittaker proposed the five-kingdom scheme most widely used today. In that system, bacteria and cyanobacteria (or blue-green bacteria) are placed in the kingdom Monera. A separate kingdom distinction is warranted because not only are bacteria and cyanobacteria true organisms composed of only one cell, but they also are the only organisms whose cells do not contain a nucleus (known as **prokaryotes**). The cells of all the other creatures that make up the other four kingdoms do possess a cell nucleus (known as **eukaryotes**).

Microbe hunters classify (group) monerans using such characteristics as cell structure, chemistry of the cell wall, and

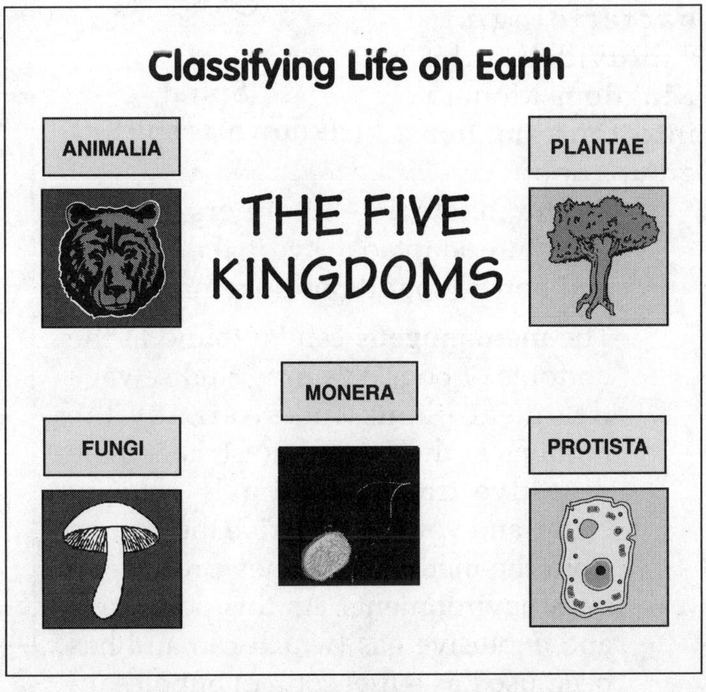

(See page 75 for full-size chart.)

methods of obtaining energy (food). Another useful classification tool is the shape of the cell. Even though there are thousands of different species of monerans, most have one of three basic shapes: round, rod, and spiral.

Another useful classification clue is the ability of the cell wall of monerans to take up certain stains (dyes). The most notable is the Gram stain. This staining method, named after Danish doctor Hans Gram, sepa-rates bacteria into two (2) groups: Gram-positive bacteria, which re-tain a stain called crystal violet, and Gram-negative bacteria, which lose the stain when treated with alcohol or acetone. Using these classifi-cation characteristics, bacteriologists subdivided the Kingdom Monera into four smaller groups or phyla:

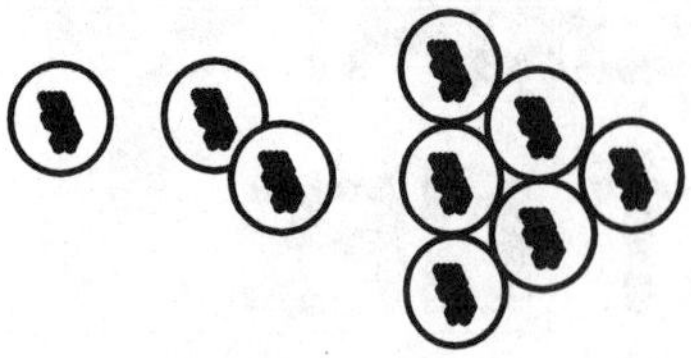

Round
(known as cocci)

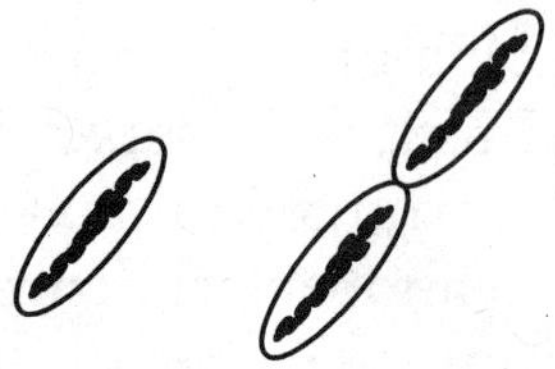

Rod
(known as bacilli)

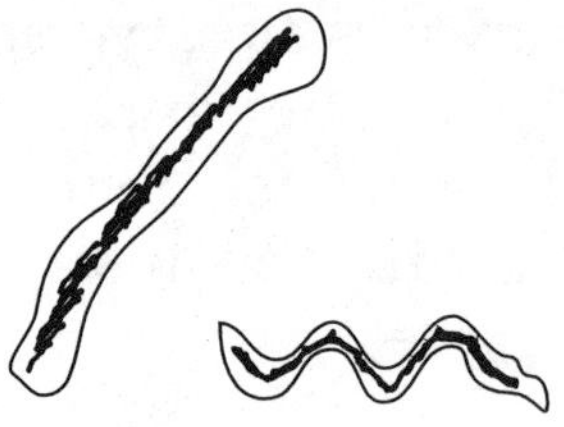

Spiral
(known as spirilli)

- **Archaebacteria**—Members of this group are adapted to live in the harshest and hottest conditions imaginable.

  The **methanogens** can be found at the bottoms of bogs, swamps, and sewage treatment plants and swarm by the millions in the warm, moist, anaerobic digestive tracts of animals—sheep, cattle, and you. They derive their name from the methane gas they produce in such environments. As this poisonous and explosive gas (which can and has been used as a fuel source) bubbles up from swamps and sewage pits, it is known as "swamp gas." When it bubbles out of animals and humans, it is known by many names but the correct term is flatulence. You may be surprised to know that the average person totes around about 2 liters (roughly 2 pints) of gas in their intestines at all times and passes bubbles of this gas an average of 14 times a day. The flammable nature of this gas has been known to cause people to explode during surgery when electrical operating equipment ignited the methane in these unfortunate people's intestines.

  The **extreme halophiles** live only in areas of high salt concentration such as the Great Salt Lake and the Dead Sea.

  The **thermoacidophiles** live in the cauldron of bubbling acidic hot springs where temperatures may reach the boiling point of water and the acidity level is nearly that of battery acid.

  Still other kinds of archaebacteria live miles below the ocean surface around warm volcanic vents, where they use the sulfur bubbling out of these vents as an energy source. Scientists think present-day archaebacteria live in environments that approximate those during the early

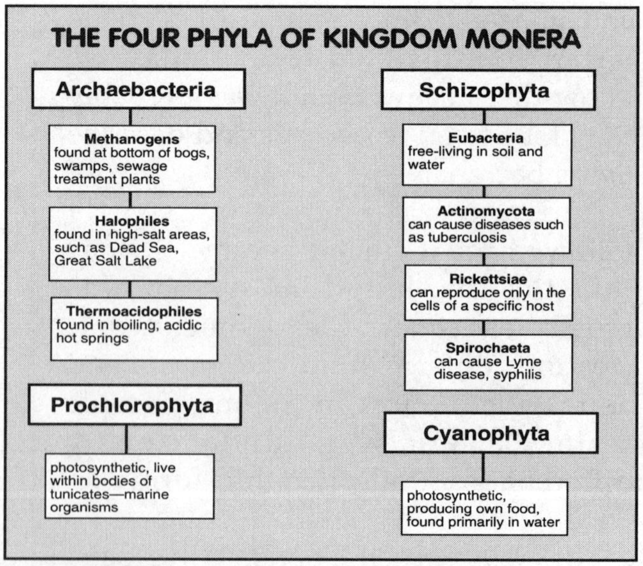

(See page 76 for full-size chart.)

days on Earth when they first appeared. Hence their name, *archae bacteria*, meaning "ancient bacteria."

- **Schizophyta**—This is the largest moneran phylum and is divided into four classes:

  Class Eubacteria contains the largest number of and many of the most familiar bacteria. Most members of the class are free-living (are not parasites and do not cause disease) soil and water bacteria.

  Class Actinomycota contains rod-shaped organisms that form colonies of branching filaments. Some actinomycotes decompose dead plants and animals, and some cause diseases such as diphtheria and tuberculosis.

  Class Rickettsiae contains parasitic bacteria that can reproduce only in the cells of a specific host. Insects often carry these bacteria and transmit them to animals. Typhus is an example of a rickettsial disease transmitted by lice.

  Class Spirochaeta contains spiral-shaped or curved bacteria. Most spirochetes use flagella (hairlike whips) to move. One species causes the sexually transmitted human disease known as syphilis. Another causes Lyme disease, an ailment similar to arthritis that is transmitted by ticks.

- **Prochlorophyta**—These strange microbes are photosynthetic but live symbiotically within the bodies of marine organisms known as tunicates.

- **Cyanophyta**—Members of this group are also photosynthetic and produce their own food. However, unlike true plants and plantlike protists, the chlorophyll of these bacteria is not organized into distinct structures known as chloroplasts but rather floats free in the cytoplasm of the cell. It is this greenish chlorophyll that gives the microbes in this phylum the name of cyanobacteria or blue-green bacteria. Cyanobacteria are found primarily in water, where they clump together in colonies of chains or clusters.

Scientific names and groupings are artificial scaffolding constructed by humans to organize and understand the world around them. In reality, nature moves without concern or respect for the definitions imposed by humans. Even so, we need such names to accurately communicate about organisms, groups of organisms, and ecological connections.

Because of their small size, lack of obvious physical characteristics, and sheer numbers, microbes are the most difficult creatures to classify. There are thousands, maybe tens of thousands of species of monerans and even professional bacteriologists find the task of identifying and naming them overwhelming. But take heart! While you may not be able to specifically identify it, your chances of finding a microbe new to biology, especially in muds and waters, are very good.

(See related student activity, pp. 60–1.)

Because of their small size, lack of obvious physical characteristics, and sheer numbers, microbes are the most difficult creatures to classify.

Chapter 2: Bacterial Structure—Staying Alive

Between the nonliving virus particles of the shadow zone and the amazingly complete cells of plants and animals, we find the simplest of life forms—monerans (bacteria).

Bacteria are at once the smallest, most abundant, oldest, and newest creatures on the planet:

- **Smallest in size**—That there are approximately 30 trillion bacteria per ounce in fertile soil and that 10,000 of them laid end to end would fit only across your fingernail are testimony to their minuteness.

(See page 77 for full-size chart.)

- **Most in number**—Bacteria are the most numerous organisms on the planet. The population of bacteria in a single pailful of fertile soil is far larger than the population of humans presently living on Earth.
- **Oldest on Earth**—Fossil bacteria dated to over 3.2 billion years old have been found in black chert, the oldest organisms of all, in South Africa.
- **Newest on Earth**—Recombinant DNA (popularly called genetic engineering) has produced new species of human-made bacteria, the newest organisms of them all.

Bacteria are prokaryotic. That is, they lack an organized nucleus. They possess DNA (genes) but the DNA strands float free in the cellular cytoplasm. This then is the defining characteristic of all monerans (bacteria). Bacteria also lack most of the internal cell parts (organelles) found in more complex cells. The cells of all other types of living things are eukaryotic, possessing an organized nucleus and all internal cell organelles (meaning "tiny organ"). If cells were vehicles, a bacteria cell would be a push cart while all other types of cells would be Indy race cars.

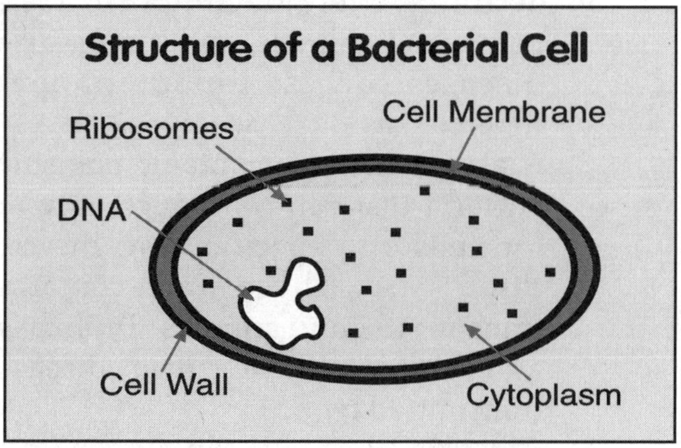

(See page 78 for full-size chart.)

All bacteria have the following structure:

- **Cell Wall**—This is a thick layer that contains the cell. In some types of bacteria it is rigid; others have flexible cell walls. Plant cells also have a cell wall, but the chemicals that make up the cell wall of bacteria are different than those composing the cell wall of plant cells.
- **Cell Membrane**—This is a thin, selectively permeable layer around the cell that allows materials to move in and out of the cell.
- **Cytoplasm**—This is a thick jellylike material that fills the cell.
- **Ribosomes**—These are organelles that synthesize (manufacture) protein.
- **DNA**—These strands, usually arranged into circular chromosomes, are the genetic material that direct the cells' life activities.

In addition, some species also form the following secondary structures:

- **Capsules**—This is a sticky layer some bacteria form over their cell wall. It helps the cell cling to surfaces. Disease-causing bacteria often have a capsule which may protect them from the host's immune system.

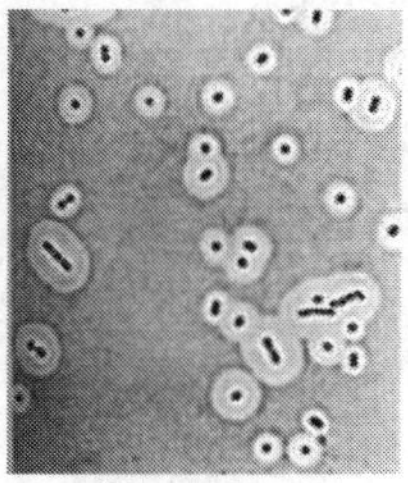

Capsules

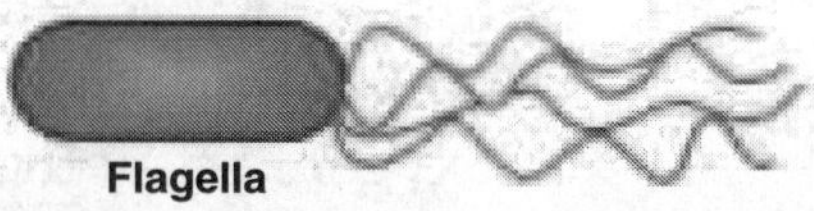

Flagella

- **Flagella**—These long hairlike projections help some bacteria move around. Flagella act like tiny propellers rotating and moving the cell.

- **Pili**—Used like grappling hooks, these short hairlike protein prongs help some types of bacteria cling to surfaces to obtain food.

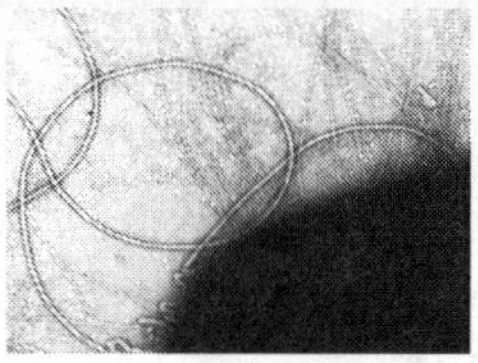

Pili

- **Endospores**—Under harsh conditions, certain bacteria can form a thick, protective wall within them around their DNA (genes). The rest of the cell may die, but the endospore will continue to survive. Endospores are resistant to high temperatures, drying, some toxic chemicals, and radiation. When the environment becomes favorable, a new cell can emerge, sometimes as long as 50 years later. Many disease-causing bacteria can form endospores. Therefore, methods to kill such spores are important in food preservation, medicine, and laboratory procedures to prevent the growth of potentially harmful organisms.

Endospore

(See related activities pp. 55, 58–61.)
(See page 79 for full-size graphics.)

Meeting Nutritional Needs

In spite of their small size and comparatively simple structure, bacteria, like any other living thing, must obtain food. Monerans can be categorized into two groups based on how they meet their nutritional needs:

- **Autotrophs**—This word literally means "self-feeding" and applies to the fact that several groups of monerans produce their own food from within through photosynthesis or chemosynthesis. The cyanobacteria (or blue-green bacteria) are one such group of photosynthetic bacteria. However, photosynthesis in bacteria differs somewhat from the same process in green plants. The chlorophyll in bacteria is chemically different from the chlorophyll in green plants; the chlorophyll in bacteria is found on folded membranes in the cytoplasm of the cell rather than in organelles called chloroplasts, as it is in green plant cells. The chemosynthetic bacteria, like photosynthetic types, make their own food from carbon dioxide. However, they do not use light as the source of energy. Instead, the reaction is driven by the energy obtained from associated chemical reactions involving inorganic molecules, such as hydrogen gas and ammonia.

Chemosynthetic Bacteria Recipes

$CO_2 + H_2$ = Food
$CO_2 + NH_3$ = Food

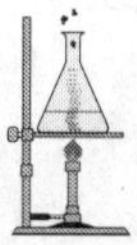

Carbon dioxide mixes with hydrogen gas or ammonia to produce food.

(See page 80 for full-size chart.)

- **Heterotrophs**—Monerans in this category must obtain their food from the environment outside them. Most heterotrophic bacteria feed on the dead remains of all other forms of life. Some live peacefully inside other creatures—warm, moist, and surrounded by the food eaten by the host. An example would be the bacteria that live in our own intestines. A few types take a more sinister approach. They live in a host, but they kill and digest the host's cells. Such parasitic bacteria are called pathogens and are those that cause the diseases we associate with bacteria.

TWO GROUPS OF MONERANS

Autotrophs

produce their own food through photosynthesis (from chlorophyll) or chemosynthesis (from carbon dioxide)

Heterotrophs

get food from their environment (by feeding on dead remains)

(Full-size chart, page 81)

In higher life forms, once food is obtained or produced, it must be broken down to release energy in a process called respiration. This process requires oxygen. Some bacteria, known as **obligate aerobes**, use this process as well. The bacteria that lives in the human lung and causes tuberculosis is an obligate aerobe. Others, such as the methanogens, find oxygen poisonous and are **obligate anaerobes**—living removed from the presence of oxygen. Still others, such as some of the bacteria in our digestive tract, can live with or without oxygen and are known as **faculative aerobes**.

Though comparatively simple in structure, a single bacterial cell amazingly undergoes all the same life processes within its tiny being that you do on a grander scale with trillions of cells arranged into a multitude of organs.

(See related student activities, pp. 62–5.)

Chapter 3: Bacterial Ecology—They're Everywhere

Virtually everywhere we look we find thriving hordes of bacteria. Why? Because the number of microhabitats available to bacteria far surpasses the number of macrohabitats in which plants and animals can live. This is due to the fact that not only are bacteria smaller, but they are also far more adaptable. Each bacterial generation is, in effect, an experiment in evolution. As such, bacteria have been able to adapt to an amazing array of different living conditions and locales.

Some flourish in boiling hot springs, others in acidic or salty "soups" that would quickly kill other living things. Bacteria have been collected with an air filter from the stratosphere and brought up from the cold silent muds of the ocean floor.

Not only do bacterial habitats include the surface area of the planet, extending more than 7 miles down into sea water and bottom ooze and up miles into the air, but they also include on and within the bodies of larger organisms. Bacteria have even traveled into space as micronauts on and in human astronauts.

(See related student activities, pp. 58–9.)

When examining the distribution of larger creatures in nature, one must consider geographical, geological, and even historical factors. Not so with bacteria. Because of their small size, bacteria are not greatly restricted by the factors that regulate the dispersal of other organisms. Bacteria move freely above, over, and through the planet in a number of ways:

- **Air Dispersal**—While it is likely that little or no growth of bacteria occurs in the air itself, every gust of wind will stir bacteria and launch them to new possibilities. Bacteria have been found in air samples from 6 miles high.

Airborne bacteria may occur individually, but they more frequently cluster together in "rafts." Even though heavier than air, the slightest turbulence or moving thermal current is enough to keep them aloft. Imagine being able to actually see the bacteria floating in the air. In a sealed room, one would begin to perceive a gentle bacterial "rain" as the air grew still, and the bacteria slowly tumbled to the floor.

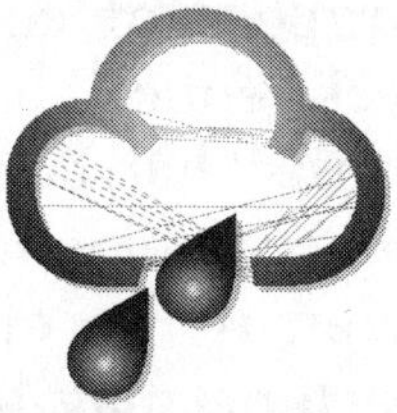

- **Water Dispersal**—Few bacteria are found—alive or dead—in open water away from sands, sediments, or the bodies of living things. The number of bacteria found in sea water is far lower than would be expected on the basis of the amount of organic matter that exists there. Further, most of the bacteria do not exist free in the water but live attached to plankton, algae, or dead or decaying matter. Thus, it appears that the bacteria found free in open water are, like those in air, merely on their way from one place to another, aquatic nomads looking for new habitats to conquer.
- **Growth Dispersal**—Growth itself leads to dispersal to new environments. Even though dispersal by growth can be over only short distances, it may still be of considerable importance, especially in the microenvironment of soil, which swarms with bacterial life.
- **Animal Dispersal**—Animals of all types are teeming with external and internal bacterial populations, especially on the skin and in the intestinal tract. The human mouth alone may harbor 50 billion bacteria.

Human activities provide some unique modes of transportation of bacteria and viruses. We constantly move exotic flowers, fruits, trees, pets, livestock, foods, and inanimate objects around the globe in a matter of mere hours. In the process, we provide access to areas that bacteria and viruses could never reach on their own.

Ignoring the geographical and geological constraints that contain other life forms, bacteria move freely up into the air, down into the inky depths, and across the face of the land. This planet with all its nooks, crannies, and crevices is their possession; they truly are everywhere.

And as in the human world of politics and societies, methods of attack and defense, acquisitions and mergers, coalitions and alliances are all found in the microbial world. Bacteria with the colorful name of **Vampirococcus** can attack, devour, and grow, killing an entire population of other bacteria. **Bdellovibro** bacteria must invade and reproduce inside other bacteria in order to survive. After digesting their host from the inside out, they reproduce and burst out, spreading and entering new hosts. Some cell biologists believe that the **mitochondria**, the "powerhouse" organelle indispensable in eukaryotic cells like our own, may have evolved at some point in the distant past from the same sort of bacterial invasion.

In the hindgut of a termite, one finds the swarms of protozoa that actually digest the wood (cellulose) eaten by the termite. Yet if one looks closer, small **spirochete** bacteria can be seen pushing the much larger protozoa around and even attaching to and feeding off the surface of the protozoa.

Though inspired by Anton van Leeuwenhoeks's discovery of tiny parasites on fleas, the words of the great satirist Jonathan Swift ring true in the hidden world of microbes as well:

So, Naturalist observe, a flea hath
smaller fleas upon his back to bite him
and the smaller fleas have smaller fleas
and so proceed ad infinitum.

(See related student activities, pp. 60–5.)

Chapter 4: Bacterial Reproduction—Be Your Own Parent

Viruses do not undergo true reproduction. Instead, they sabotage cells to replicate new viral particles. However, being true living cells, bacteria must reproduce themselves. And reproduce they do. Each bacterial cell duplicates its own DNA and then simply divides.

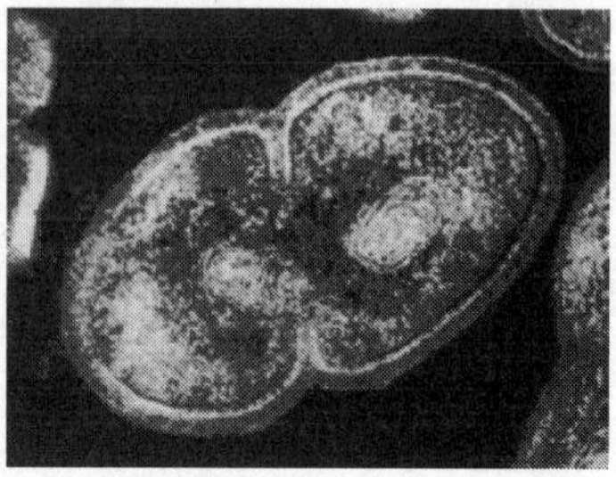

This process of reproduction through direct division is called **binary fission**. Under favorable conditions, bacteria can divide every 15 to 20 minutes. Theoretically, if that rate could be maintained, in 48 hours the bacteria would reach a solid mass 4,000 times bigger than Earth. Fortunately, lack of food and space, accumulation of wastes, and predation prevent the planet from being paved over with bacteria.

Bacteria are usually asexual reproducers. However, under certain conditions, some bacteria engage in a primitive form of sexual reproduction (where sex is defined simply as biological shorthand for gene exchange) called **conjugation**.

During conjugation, a long bridge of protein forms between and connects two bacterial cells. Part of the genetic information (often in the form of circular bits of DNA called **plasmids**) from one cell, called the **donor**, is transferred to the other cell, called the **recipient**, through this bridge. When the process of conjugation is complete, the recipient cell has a different set of genes from those it had before conjugation occurred. This new combination of genes helps to ensure that even if the environment changes, a few bacteria may have the right combination of genes to survive. Such was the case with Donna (p. 36). Once the flesh-eating bacteria secured a solid footing, it survived her body's fiercest antibody assaults.

This sort of gene transfer has been called "horizontal" and contrasts with the "vertical" reproduction of larger, sexually reproducing organisms. Larger organisms must produce offspring in order for genetic recombination to occur. However, a bacterium receives genes via a partner and then proceeds to clone (asexually reproduce) itself. The bacterium directly inherits and maintains new traits with impressive fidelity. This ability to adapt quickly to changing environments and to pass on that ability in unmixed form may be a major mechanism of planetary environmental control and certainly lies at the heart of the developing genetic engineering industry.

(See related student activity, pp. 65–71.)

Chapter 5: Bacterial Impact—Microbes Rule

We turn a condescending eye on bacteria, yet life as we know it would be impossible without these lowly bacteria.

How do bacteria affect the existence of all the other crew members of Spaceship Earth, especially humans?

- **as atmospheric converters**—Ancient photosynthetic and chemosynthetic bacteria (similar to present-day cyanobacteria and methanogens) began the process of converting the poisonous methane-ammonia atmosphere of Earth to the benign life-supporting nitrogen-oxygen mix of today's air.
- **as recyclers**—Decay bacteria are critical in freeing atoms, molecules, and minerals from the bodies of dead organisms so they can be recycled back into the air, water, and soil and, ultimately, into the bodies of living organisms. If decay bacteria were to go on strike, within several years there would be a layer of

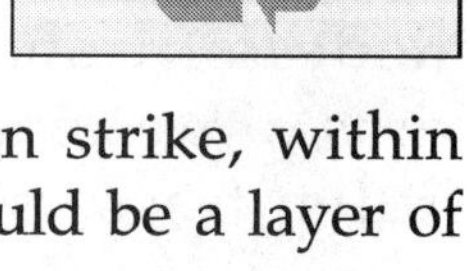

dead material at least 5 feet thick covering the surface of the land and water. New life forms would be limited if not impossible under such conditions.

- **as rejuvenators**—The process of decay renews the minerals and elements in the soil withdrawn by the multitude of plants and animals living on and in it. All plants have to have nitrogen. Yet despite the simple fact that nitrogen accounts for nearly 80% of our planet's atmosphere, no plant by itself is able to obtain this important element directly from the air. Plants depend instead on certain bacteria in the soil and in their roots to "fix" this atmospheric nitrogen into a usable form.

Flower: Bacteria-rejuvenated

- **as death dealers**—Parasitic bacteria are the agents of many human, animal, and plant diseases. They inflict plants with wilt, gall, spot, and canker; bedevil animals with blackleg, anthrax, Rocky Mountain spotted fever, and fowl cholera; and plague humans with diphtheria, tuberculosis, typhoid fever, tetanus, syphilis, food poisoning, Strep throat, cholera, bacterial pneumonia, leprosy, and bubonic plague.
- **as life protectors**—Bacteria can cure as well as kill. Harmless heat-killed bacteria are injected into our bloodstreams. Called "vaccines," these complex substances induce the body's immune system to produce proteins called antibodies that destroy the target disease-causing bacteria. Furthermore, some bacteria produce antibiotics—natural chemicals that kill or prevent the growth of other, more harmful bacteria.
- **as food spoilers**—Decay bacteria play no favorites and often turn their talents toward decaying our food and crops. It has been estimated that bacteria and fungi ruin 1/3 or more of the world's foodstuffs every year, and many billions of dollars are spent yearly worldwide in an often futile attempt to prevent or at least delay this process.

Milk and cheese: bacteria-produced

- **as food producers**—Bacteria are essential in the production and flavoring of such foods as cheese, buttermilk, yogurt, sauerkraut, and vinegar, to name a few. In fact, bacteria themselves are food for a host of small creatures such as protozoa and perhaps, in the future, humans. With the growing shortage of food worldwide, bacteria and some algaes show some promise for being efficient and highly nutritious microcrops.
- **as food digesters**—Bacteria coat the outside and fill up much of the inside of other living things. Trillions of them live on our skin, in our bowels, even on the surface of our eyes. Those in our digestive tract help break down our food-producing vitamins and minerals in the process. In humans, 30% of the weight of fecal matter is due to bacteria, and we humans void about 10^{11} (100,000,000,000) bacterial cells per day.
- **as chemical plants**—Bacteria are indispensable to certain industries. Bacteria produce acetone and butanol; a variety of organic acids necessary in foods, pharmaceuticals, and paints; and solvents used in the manufacture of plastics. Lactic acid produced by bacteria is used to preserve food, to finish silk rayon fabrics, to make acrylic resins, and to do a great many other things. Scientists are constantly searching for new bacteria (especially in soil) that might be tapped for marvelous

new chemicals. For example, in a corn field, a graduate student found a bacterial ice-forming protein that is now used by hundreds of ski resorts to make artificial snow, a clump of soil from a Japanese golf course harbored a bacterial cure for a parasitic infection that attacks livestock, and a microbial molecule from an Indonesian temple is now used by soft-drink suppliers to turn starch into sugar.

- **as microbe miners**—As our natural resources diminish, we must today depend on mining ores that at one period were dismissed as unworkable because of inaccessibility, cost factors, and low metal content. The search for efficient, cost-effective means to extract metals has led to the use of bacteria in order to recover copper, nickel, and even uranium. Such bacteria oxidize iron, copper, and other metal sulfides by growing in very acidic solutions. The metals are thus leached out from the rocks and recovered. A number of countries have major microbial mining processes in operation.

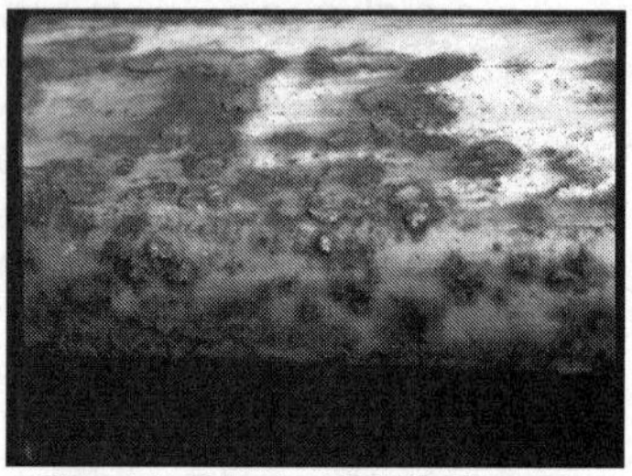

Rust: bacteria-riddled iron

- **as toxic avengers**—Bacteria that can slurp down an alphabet soup of hazardous materials offer a permanent, cost-effective solution to a variety of pollution problems. Using bacteria in this way is referred to as biological remediation, and its uses and benefits are rapidly becoming evident. Imagine using bacteria to digest oil; mop up wood preservatives; break down DDT, TNT, and PCBs; destroy toxic petrochemicals (including benzene, xylene, toluene) and gasoline; chew up some components of solid rocket fuels used in nuclear weapons that are to be dismantled; and even help in the clean-up of the radioactive uranium used in these kinds of weapons. The potential for bio-remediation is definitely huge. More than 30,000 hazardous waste sites have been identified in the U.S. alone.
- **as genetic engineers**—The transfer of DNA and RNA from one cell to another—creating new life forms that can be tapped for medical, agricultural, pollution control, and industrial purposes, has become a highly developed science with important ramifications. As we might suspect, the workhorse of this new technology is bacteria. Because they have relatively few genes that can be easily altered, bacteria can be made to produce anything from human insulin to superplagues for germ warfare. The brave new world of genetic engineering and biotechnology holds both promises and possibly pitfalls and is sending a shock wave of concern through society.

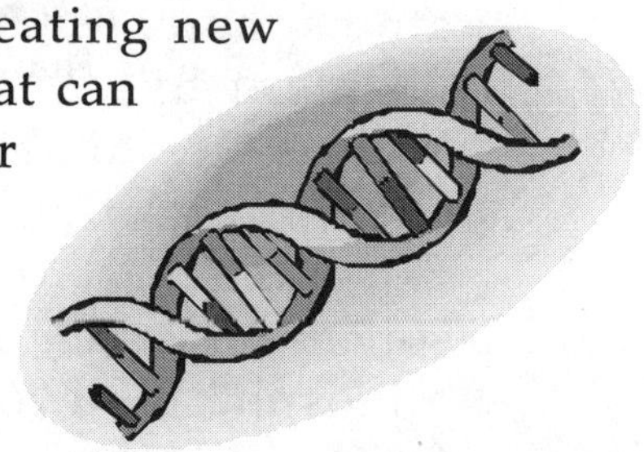

Bacteria, those most ancient of life forms, have ruled and will continue to rule the earth and hold dominion over the other living creatures within their realm.

Teacher Procedures/Information

Bacteria Concept Map

Objective: Students will complete a concept map to demonstrate a scheme for organizing the characteristics of monerans.

Student Activity is on page 55.

Answers appear on page 72.

Bacteria Brain Bogglers

Objective: Students will analyze and solve two sets of problems:

A. Given bacterial growth rates, students will project future rates and use inductive reasoning to explain causes of rate change.

B. Given information on two patients, students will diagnose the cause of illness as induced by either virus or bacterium.

Student Activities are on pages 56–7.

Answers are on page 73.

A Bacteria 'Trap'

Objective: Students will collect and grow bacteria as a basis for observation and correlation of high bacterial count to type of environment.

Student Activities are on pages 58–9.

Answers are on page 73.

A. Background

Bacteria are found in large numbers practically everywhere. Demonstrate this to your students by showing them a spoonful of soil and have them speculate on how many individual bacteria are present in that much soil. (Write all student responses on the board.) Then ask students to draw a square centimeter on a piece of paper. Ask them to predict how many bacteria are present on a square centimeter of human skin. Finally, hold up a gram weight and ask students to guess how many bacteria are in a gram of digesting food in a person's intestine. Amazingly, there are 10 billion bacteria in a spoonful of soil, 1 billion per square centimeter of skin, and 10 billion per gram of digesting food in the human intestine.

B. Materials

- 1 petri dish of nutrient agar per lab group
- 1 roll (or at least 3 pieces) of transparent tape per lab group
- 1 marker per lab group

C. Procedure

1. Prepare the dishes of nutrient agar before class. The suppliers listed in the Appendix at the back of this book will sell dishes containing nutrient agar, the dishes and bottled nutrient agar separately, or the dishes and the ingredients with instructions to make your own nutrient agar.
2. Students should divide the bottom of the dish into 4 sections by marking the underside. Why the bottom? If students divide the lid, the lid may get turned, resulting in confusion about which section is which. Some mark that identifies the dish should be placed on the lid.
3. Someone in each lab group should gently touch Section 1 in the dish. Students should try to press only hard enough to leave a fingerprint on the agar. If a student pushes too hard and breaks the agar, no problem will result, but the bacterial growth will be hard to see down in the broken piece of agar.
4. If you do not have enough rolls of tape to go around, have one member of each lab group put a long piece of tape on three of a student's fingers instead and send him or her out.
5. Anywhere and everywhere in school should be fair game as long as students

Teacher Procedures/Information

A Bacteria 'Trap' (cont.)

are not disrupting other classes or normal school function. Encourage students to use their imagination as to what surfaces or objects they test. Instruct them to keep track of where they went and what they tested in the table on their activity pages.

6. When students return, they should place their dishes upside down in a safe place where the bacteria can grow for several days without being disturbed. If you have access to an incubator, set it up and turn it on about 24 hours before you conduct this activity. The best temperature for growing most bacteria is 37°C, and you need to get the incubator stabilized at this temperature before you begin. With an incubator, you should have results in 24 to 36 hours. Leaving the plates to grow at room temperature will certainly work, but it will take longer (at least 2 to 3 days) to observe any growth.

7. Present questions and data and let students work individually or in groups to answer them. You may want to have a class discussion of the thinking involved.

8. *Safety Note*—Chances are slim that students will trap any disease-causing bacteria and if they do, very slim that such bacteria could grow in only nutrient agar. However, be on the safe side and instruct students not to remove the lids or touch anything inside their dishes after the bacteria have had time to generate visible growth. Once the activity is complete, sterilize each dish by pouring full-strength bleach into it and letting it sit for 1 hour. If using plastic dishes, place the sterilized dishes into a strong plastic bag and dispose. If using glass dishes, dump the sterilized contents of each dish into a strong plastic bag, dispose of the bag, and wash dishes well in strong detergent.

Observing Bacteria

Objective: Students will prepare and view a specimen under a microscope in order to observe and describe bacteria.

Student Activity is on pages 60–1.

Answers will vary with student experience. Suggested answers are on page 73.

A. Background

This activity allows students to see actual bacteria and learn how bacteriologists prepare and stain bacteria slides for viewing (optional).

B. Materials

- 1 microscope per group of 2 or 3 students if possible. The microscope must be capable of at least 100X magnification
- Part I: 1 commercially prepared slide of bacteria per group
- Part II: 1 slide, cover slip a group
- a source of bacteria

A number of ways of collecting or growing bacteria that are safe to use in this activity, including the following:

1. Have students scrape tartar from their own teeth with a toothpick.
2. Grow some by placing a small handful of dried dog or cat food into a glass jar about 1/2 to 3/4 full of water. Place the jar in a warm, dark location. In about 48 hours, the water should be cloudy with bacteria.
3. You can also grow your own bacteria by placing a small jar of whole milk in a warm,

Teacher Procedures/Information

Observing Bacteria (cont.)

dark location. In several days, it will be teeming with bacteria.

- a way to dry the students' slides
- bacterial stain. Either methylene blue or crystal violet stain will give the desired results. Bacterial stains may be ordered from the suppliers listed in the Appendix.
- plastic squeeze bottles containing water Small beakers of water with droppers will also work.

C. Procedure: Part I (Optional)

Part I of this activity is optional. If slides are available, doing Part I will allow students to see bacteria and know what to look for when they make their own slides. Bacteria slides may be purchased from the suppliers listed in the Appendix in the back of this book. Purchase finished stains, not stain powders.

D. Procedure: Part II

1. The bacteria source:
   a. If you want students to view bacteria from their own bodies, have them scrape some of the tartar from their teeth with a toothpick and mix it into a drop of distilled water on the slide. Tap water will suffice but may leave mineral residue that could obscure the bacteria when the slide is dried. Instruct students to mix the tartar thoroughly with the water. This spreads out the bacteria and makes them easier to see as individuals.
   b. If you go with growing bacteria in milk or pet food in water, just have the students use a dropper to transfer some of the liquid to their slide. Be prepared for some strange, strong odors emanating from the decaying milk and/or pet food cultures.
2. After dropping bacteria on their slides, students need to dry their slides. Ideally, a Bunsen burner works best; but slides can be dried over alcohol lamps, candles if necessary, and perhaps even hair dryers. Remind students to follow all safety procedures if working around open flames. Drying the slide accomplishes two things: (1) it kills the bacteria, rendering them harmless and (2) it sticks the bacteria to the slide so they aren't washed off when the stain is removed.
3. Students should cover the dry crusty area that remains on their slide after drying with bacterial stain. Bacterial stains can permanently stain clothes, so students should take care when working with them.
   To prevent stain from getting on things other than the slide, consider having students put their slides in sinks if available or in small, flat dishes, then adding the stain. The stain should remain on the slide 3–5 minutes. This gives the stain time to penetrate the tough cell wall of the bacteria.
4. The stain needs to be rinsed off the slide. Have the students gently rinse the stain into a sink or jar using squeeze bottles or droppers full of water. Emphasize *gently* or stain may be blown on papers, skin, and clothes.
5. Have students use paper towels to blot, not rub, their slides dry. If they rub, they will wipe off the bacteria they have worked so hard to prepare. The slides are now ready for viewing.

Build a Bacterial 'City'

Objective: After building and observing a soil column for several weeks, students will distinguish different types of bacteria and correlate them to environmental conditions.

Student Activity is on pages 62–4.

Answers will vary according to student experience. Some answers are on page 74.

A. Background

Russian scientist Serge Winogradsky

Teacher Procedures/Information

Build a Bacterial 'City' (cont.)

opened the field of soil microbiology in the 1800s. During the course of his work, he developed soil column devices. Building one of these devices will allow students to visit, see, and appreciate the worlds of soil, muds, and sediments where life may have originated more than 3 billion years ago, all without the necessity of a microscope.

B. Materials

- 1 translucent clear container per group
- 1 or 2 large buckets of soil, mud, or sand
- 1 roll of clear plastic wrap

C. Procedure

1. Any clear translucent container will do. Plastic soft-drink containers are ideal. Although these containers may allow enough oxygen to diffuse and prevent the formation of certain anaerobic bacteria, this drawback is far outweighed by their availability and ease of use of the containers. They must be at least six inches tall in order for anaerobic layers to form.
2. To speed construction and to prevent student accidents, ahead of time you should cut the tops off the containers approximately at the level where the bottle widens below the neck. Save the tops to use as funnels to fill the bottles.
3. You can collect or have students bring in samples of mud or soil to use. Soils (the gooier and smellier the better) from shallow pond bottoms, mudflats, or beachfronts will form layers more rapidly because they already contain large numbers of each kind of bacteria. You might also collect garden or forest soil.
4. Mix your soil, mud, and goo with enough water to make it about the consistency of heavy cream.
5. You need a quick food source for the organisms in the column. Hard-boiled egg yolks or the entire raw egg provide an excellent sulfur and protein source. You can also use cheese or powdered calcium sulphate. For a long-term carbon source to provide carbon dioxide, use shredded newspaper. You can also sprinkle in some chalk dust or plaster of Paris.
6. Have students fill their containers in a well ventilated area. Open windows and even a fan or two may necessary because the odors that occur may be nearly overwhelming. Students may complain; but when the colors appear, the strange smells will be long forgotten.
7. As students fill their bottles, they should tap out air bubbles. Any air bubbles in the column will trap oxygen and prevent that area from becoming anaerobic.

BUILDING A SOIL COLUMN

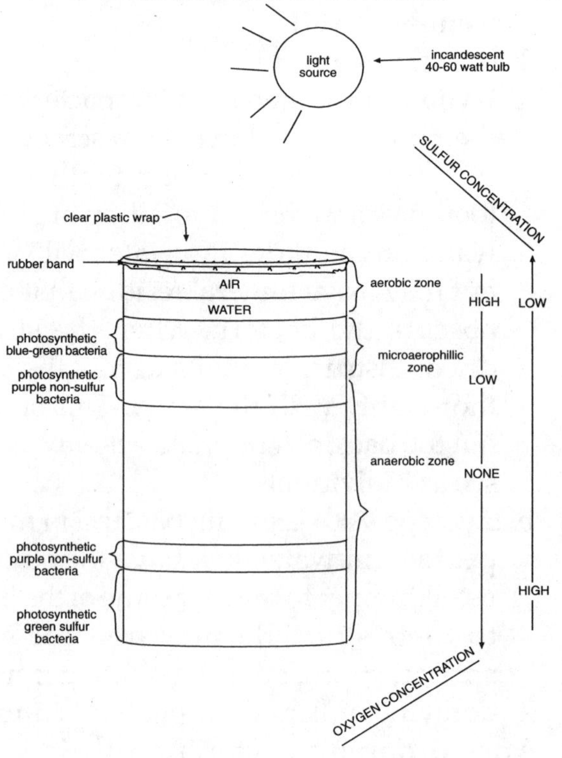

Teacher Procedures/Information

Build a Bacterial 'City' (cont.)

8. Have students either place their completed bottles in north-facing windows or light them with incandescent bulbs. One incandescent light fixture will be enough for five to ten bottles. Do not have the lights too close because heat will kill some of the bacteria. One area of a room can be set up for all the containers and become part of a permanent and ongoing display.
9. Columns can take between three and six weeks to become well established with the different layers or regions of colored bacteria. You may not get all the layers, and they may be more mixed together than in straight lines, but you will get to see different kinds of bacteria without a microscope.
10. Do not allow the columns to dry out. Top with water as needed.
11. Have students take measurements and observations of their "cities" regularly (twice a week) and record this information on their observation sheet (page 64).

Make Your Own Yogurt

Objective: Students will make yogurt to demonstrate speed of bacterial growth; they will compare a control can (plain) and an experimental can (added yogurt culture) of condensed milk for several days.

Student Activity is on page 65.

Answers will vary with student experience.

A. Background

This activity will illustrate both the practical uses we make of bacteria and the amazing speed with which bacteria can proliferate. Furthermore, making food in the classroom involves the class in a fun hands-on exploration that shows again how science is relevant to daily life.

Make Your Own Yogurt (cont.)

B. Materials

- 2 cans of condensed milk per student or group
- distilled or sterile water
- active yogurt culture
- a way of keeping the yogurt cultures warm

C. Procedure

1. Use distilled or sterile water to top off each can. Tap water may contain bacteria that will overpower the yogurt bacteria, resulting in a smelly mess rather than yogurt.
2. Setting up two cans is a good way to teach or review the concept of experimental vs. control.
3. You will need to supply yogurt containing *active* cultures. Read the labels of different yogurts at your local market carefully; some contain active cultures and some do not.
4. Keep both the test can and the control can in a warm place. If you have an incubator, set it to a temperature of about 100° F. If no incubator is available, you might set the cans on an electric warming / serving tray, you could keep them in a warm water bath, or you might set them close to an incandescent lamp (use a thermometer to check the temperature of both cans).
5. Your yogurt should "harden" within one or two days and be ready (safe) to eat after two or three days. Yogurt produced in this way will be very flat tasting, so bring in some pieces of fresh fruit to liven up the taste when you eat it.

A War on Bacteria

Objective: To analyze the effectiveness of some chemicals advertised to kill bacteria.

Teacher Procedures/Information

A War on Bacteria (cont.)

Student Activities are on pp. 66–70.
Answers will vary according to student experience. Some answers appear on page 74.

A. Background

Bacteria were first seen in the 1600s; and by the late 1700s and early 1800s, scientists had made the connection between certain bacteria and disease, decay, and death. Since then, humans have been formulating strategies and perfecting chemicals to kill and contain bacteria.

In this activity, students will test the effectiveness of some of the chemicals humans have discovered or formulated to kill or control bacteria.

B. Materials

- 1 broth (liquid) culture each of Escherichia coli and Bacillus cereus. One culture of each should provide enough bacteria for both parts of this activity. Each culture should be in a tube with a screw cap lid.

for Part I:

- 1 marker per group
- 1 dish of nutrient agar per group
- 3 or 4 different types of antibiotic sensitivity discs
- cotton swabs
- a small open container of bleach

for Part II:

- 1 marker per group
- 1 dish of nutrient agar per group
- various household chemicals advertised to kill bacteria
- cotton swabs
- forceps: 1 pair for each type of chemical to be tested

C. Procedure: Part I

Antibiotics were known by their activities long before they were given the name by which we now know them. Many years ago the Chinese controlled boils with moldy soybean curd and treated foot infections by wearing sandals furry with mold growth.

In 1929, Alexander Fleming noticed that an agar plate inoculated with Staphylococcus aureus (a common skin bacteria) had become contaminated with a mold. He was puzzled by a clear zone around the mold and postulated (correctly) that the mold was producing chemicals that were killing the bacteria. By serendipity (lucky accident), the "magic bullet"—a chemical substance that would selectively kill bacteria inside humans and animals—was revealed. The mold that accidentally led to what was perhaps the greatest medical discovery of all time was called Penicillium notatum; hence, the name penicillin for the antibiotic derived from it.

Alexander Fleming saved a sample of the original mold. He later gave it to a colleague. That colleague maintained the culture but died in 1970. His son recently auctioned off the sample to the Science Museum of London for a large sum of money.

1. Each group will need a dish of nutrient agar. You may purchase them already prepared or the ingredients to prepare them from any of the suppliers listed in the Appendix.
2. Any of the suppliers listed in the Appendix can also supply antibiotic sensitivity discs. These are heavy paper circles that have been soaked in antibiotic solution. Purchase 3 or 4 different types. I suggest penicillin, ampicillin, tetracycline, and erythromycin or streptomycin.
3. Students should use the marker to divide the bottom of their dish into the same number of sections as you have types of antibiotic discs. The discs are usually

Teacher Procedures/Information

A War on Bacteria (cont.)

identified by initials on each disc. If not, assign a certain antibiotic to each section of each dish.

4. You will need a tube of broth culture of each type of bacteria, as listed in the Materials section. You may purchase these cultures from any of the suppliers listed in the Appendix or perhaps you could get them free, as I do from the microbiology department of a local college. Cultures should grow at least 48 hours before use.
5. You should now swab, not pour, some of the liquid bacteria cultures on the agar in each group's dish. I suggest you put one type of bacteria on half the dishes and the other type on the remainder of the dishes. Open the lid of the broth culture tube, dip in a cotton swab, close the cap of the tube and open the lid of the dish. Lightly rub the cotton swab across the entire surface of the agar in the dish. You want to just lightly streak the bacteria on the surface of the agar because too much liquid will prevent the antibiotic discs from sticking to the surface of the agar. Replace the lid of the dish as soon as possible. Place the used cotton swab into a container of bleach to sterilize it and repeat this procedure with the remainder of the dishes.
6. Now add antibiotic discs. These discs usually come in a dispenser. Go from group to group and apply the discs to the surface of the agar. Students should use forceps to lightly press each disc firmly against the agar. (If the discs move around, you will not get neat, clear kill zones around the discs).
7. Students should put their forceps point down into a container of bleach to sterilize them.
8. Each dish should be placed upside down in a warm spot. Results will show up in 24 to 48 hours (perhaps longer if dishes are kept only at room temperature).
9. You and all students should now thoroughly wash your hands. All tables and countertops used during this activity should also be washed down with a 10% solution of bleach or with a full-strength cleaner such as Lysol.
10. Once the activity is complete, sterilize each dish by pouring full-strength bleach into it and let it sit for 1 hour. If using plastic dishes, place the sterilized dishes into a strong plastic bag and dispose of safely. If using glass dishes, dump the sterilized contents of each dish into a strong plastic bag. Dispose of the bag and thoroughly wash the dishes in strong detergent.

E. Procedure—Part II

Humans have concocted an amazing array of chemicals to kill or contain bacteria in practically all aspects of our daily lives. But are these chemicals as effective as we think?

1. You will need a selection of common household chemicals gathered before beginning this part. You can gather these in your local market, or it may be more interesting to have students bring samples from home. Look for key words on the labels of these chemicals such as germ-killer, disinfectant, antimicrobial, or bactericide to determine their applicability for testing. The wider the variety of materials, the more interesting the activity.
2. Before class, place a small amount of each sample in a small open container (such as a beaker) on a piece of paper identifying the chemical. Place a pair of forceps by each sample container.

Teacher Procedures/Information

A War on Bacteria (cont.)

3. Use a paper punch to make your own paper discs. Use filter paper or white construction paper. Place a supply of these blank discs in a flat open dish beside each chemical sample container.
4. Each group will need a dish of nutrient agar. Students should divide the bottom of the dish into 5 or 6 sections depending on the number of chemicals available for testing. Crowding too many discs into one dish will cause the clear zones to blend together, obscuring the results.
5. Swab bacteria on each group's dish using the same procedure explained in Part I. If mouthwashes are to be tested, do not swab the sections that will hold the mouthwash paper discs. The bacteria types you have are intestinal bacteria and decay bacteria, neither of which is a fair test for the mouthwash. Instead, have students rub a clean cotton swab over their teeth and tongue and then rub this swab lightly over the agar in the mouthwash section of the dish. This will transfer mouth bacteria to that section—a fairer test for the mouthwash.
6. One group member should hold the closed dish while another group member uses the forceps at each station to pick up one of the blank paper discs, dip the paper disc into the chemical, and then wipe the disc on the lip of the container several times. Wiping removes excess chemical, thus preventing chemicals from running together in the dish.
7. The group member holding the dish should now remove the lid of the dish so the other member can place the soaked paper disc on the appropriate section of agar. Each soaked paper disc should be lightly pressed into the agar. The lid of the dish should be replaced immediately and the forceps again placed next to the chemical container. Instruct students not to mix up the forceps or they will not be sure which chemical they are testing in each section.
8. Students should repeat this process for the remainder of the chemicals they wish to test.
9. Once each dish is complete, it should be kept upside down in a warm place.
10. At the conclusion of setting up this part of the activity, place all forceps point down in a small container of bleach. Teacher and students should thoroughly wash their hands, and tables should be disinfected with 10% bleach or full-strength cleaner. Dispose of chemical samples by pouring them down a drain. ***Safety Note***—never mix an acid such as a toilet bowl cleaner or hard-water scale remover with bleach. Such a combination can give off deadly chlorine gas.
11. Once the activity is complete, sterilize each dish by pouring full-strength bleach into it and letting it sit for 1 hour. If using plastic dishes, place the sterilized dishes into a strong plastic bag and dispose of safely. If using glass dishes, dump the sterilized contents of each dish into a strong plastic bag. Dispose of the bag and thoroughly wash the dishes in strong detergent.

Putting It on Paper

Objective: Students will use creative writing or painting to demonstrate knowledge in one of three areas:

A. the effects of bacteria on humans.
B. bacterial characteristics and habitats.
C. the benefits or drawbacks of microbes to humans.

Student Activities are on page 71.
Answers will vary with student experience.

Bacteria Concept Map

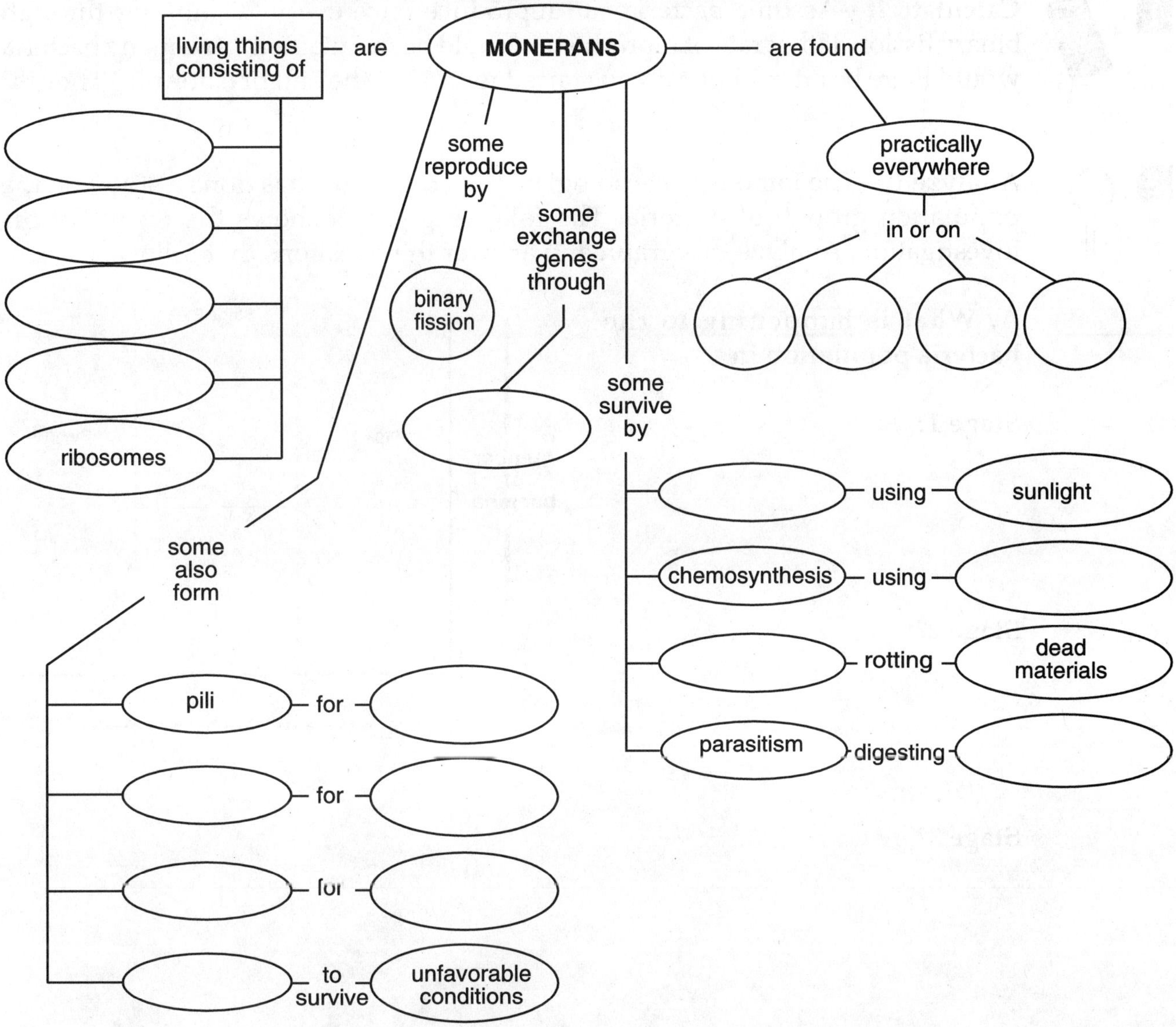

Word List

a cell wall
a cell membrane
air
attachment
binary fission
chemicals
conjugation
cytoplasm
dead materials
decay
DNA (genes)
flagella
live cells
living things
movement
photosynthesis
protection
ribosomes
slime capsule
soil
spores
unfavorable conditions
water

Bacteria Brain Bogglers

1 

Calculate It—Assume bacteria can reproduce once every 20 minutes through binary fission. If that rate of reproduction could be maintained, how many bacteria would there be after 1 hour if you started with 1? at the end of 6 hours? 12 hours?

2 

Analyze It—The famous bacteriologist Dr. Bart Bacillus has done a study on the population growth of bacteria. The following graph shows the results of his investigation. Analyze this graph and answer the questions that follow.

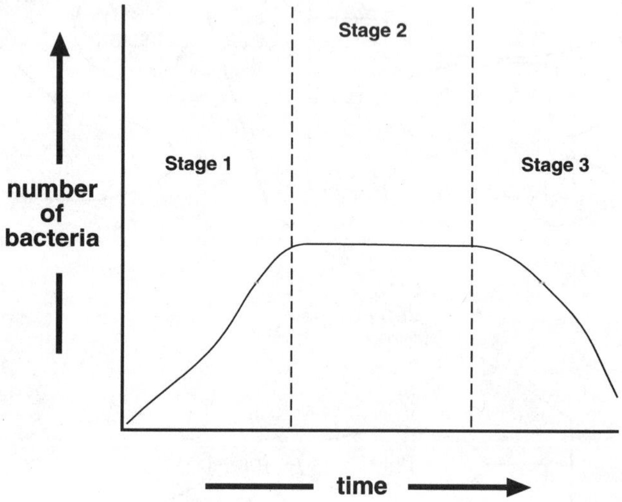

A. What is happening to the bacteria population in

Stage 1?

Stage 2?

Stage 3?

B. Why might the population have leveled off in Stage 2?

C. Why might the population have declined in Stage 3?

Bacteria Brain Bogglers

Continued

3 **Diagnose It**—Imagine you are an emergency room physician. You have two patients you suspect are suffering from food poisoning. Here are the details concerning each patient:

Patient 1—He reports that he ate his last meal about 6 P.M. Although he felt fine the next morning, the patient became very sick at work. After suffering severe abdominal pains and vomiting, the patient returned home. The patient also had a fever, chills, and severe diarrhea. He still felt sick the next day and did not fully recover for several days.

Patient 2—This patient shows signs of recovery and feels well enough to go home. You discover that the last time she ate was about 6 P.M. the night before. While watching television later that evening, the patient became ill and had extremely severe abdominal cramps. The patient had a mild case of diarrhea that has subsided. There was no fever.

You have the following facts to work with:

A. Every year thousands of cases of bacterial food poisoning are reported.

B. Two types of bacteria cause food poisoning: staphylococci and salmonella.

C. Staphylococci produce a toxin (poison) that is secreted into the food source as the bacteria multiply. Once a person eats this tainted food, the toxin is carried throughout his/her body by the bloodstream. Within a few hours, the toxin will cause symptoms that include diarrhea, vomiting, nausea, and abdominal cramps. Fortunately, recovery is usually complete 24 to 48 hours after the onset of symptoms.

D. Salmonella also produce a toxin, but it is released only when the bacteria burst. Because of this, it takes longer for a person to feel the effects of salmonella, often 12 hours or more. Salmonella infections almost always cause diarrhea. And they also generally result in fever, chills, frequent vomiting, and abdominal pain. It may also take a patient quite a bit longer to recover from a case of salmonella food poisoning.

With these facts in hand you are now ready to make your diagnosis. Which type of bacteria caused the food poisoning suffered by each patient? Support your diagnoses based on the facts.

A Bacteria 'Trap'

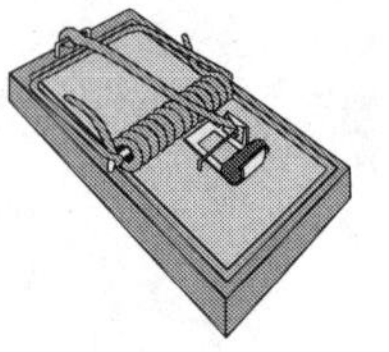

Bacteria are found practically everywhere on Earth—in air, water, and soil—and on and in the bodies of living things. But where are they found in our school? The purpose of this activity is to go out into our school and collect some bacteria on a specially prepared dish. We will do this in order to learn where bacteria live and something about growing them under laboratory conditions.

Procedure

1. Get one bacteria "trap" per lab group. Your "trap" is a special kind of dish called a petri dish. The dish is filled with a gelatin-like food substance for bacteria called nutrient agar.

2. Using a marking pen, divide the underside of the dish into four equal parts and number each part. Put your lab group name, initials, or some symbol on the lid of the dish so you can identify it later. Why is it important to mark off the bottom of the dish into sections rather than the lid?

3. Open the lid. One member of your lab group should ***gently*** touch the surface of the agar in section 1 of the dish with the tip of a finger. Don't push so hard that you break the agar.

4. Now get a roll of transparent tape. You will use this tape to collect bacteria and transfer them into your dish. Do not disturb other classes while doing this. Touch the sticky side of the tape to a surface that you think might contain a good sample of bacteria. Then open your dish and touch the sticky side of the tape with bacteria now on it to the nutrient agar in one section of the dish. ***Gently*** rub the tape on the agar to transfer the bacteria. Use a different piece of tape for each surface you test. Use the data table to keep track of what surfaces or objects you tested.

Section of Dish	**Source of Bacteria**
1	**finger**
2	
3	
4	

5. When you return, place your dish upside down in the area designated by your teacher.

6. In a few days, we will examine the dishes for bacterial growth. There is little chance that disease-causing bacteria will grow on the type of food in your dish but for safety's sake, do not remove the lid of the dish and certainly do not touch anything inside the dish.

A Bacteria 'Trap'

Continued

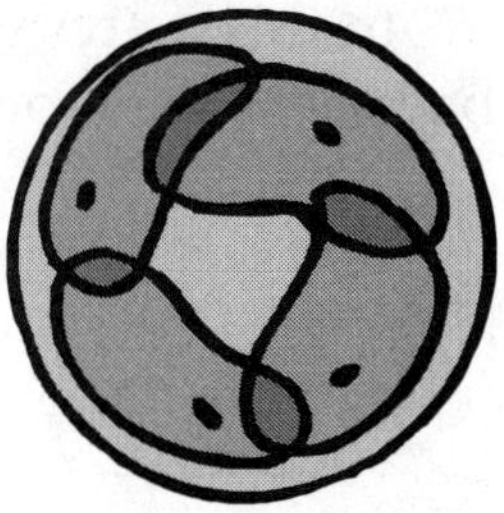

1 What was the purpose of the nutrient agar?

2 What was the purpose of the tape?

3 Why did someone in each group touch Section 1?

4 Why were you instructed not to remove the lid of your dish after bacteria growth appeared inside?

5 Why were the plates placed upside down while we waited for bacterial growth to occur?

6 The growths you see on your agar are called bacterial colonies. What is a bacterial colony?

7 What areas of school or what objects seem to have the highest counts of bacteria? Predict why there are so many bacteria in these places or on these objects.

Observing Bacteria

In this activity, we wish to observe actual bacteria and learn how bacteriologists prepare bacteria slides for viewing.

Procedure

Part I: Prepared Bacteria Slides

If slides are available, examine commercially prepared and stained slides of bacteria. Use these slides to familiarize yourself with the different sizes and shapes of bacteria before you make your own slide.

Procedure

Part II: Prepare Your Own Bacteria Slide

1. Using a damp paper towel, thoroughly clean a microscope slide.
2. Place a small drop of liquid containing bacteria on the slide. The teacher will explain where the bacteria came from and any special instructions for transferring them to the slide. Place a cover slip on the liquid. Observe under low power, then up to high power of your microscope. Hint: low light will make the bacteria stand out more clearly, so adjust your scope to allow the least amount of light to pass up through the slide.
3. Rinse off the slide and cover slip and dry them with a towel. Put the cover slip away; you will not need it for this part.
4. Again place a drop of the bacteria liquid on your slide.
5. Now slowly dry the slide over a flame. Hold onto one end of the slide and pass it back and forth through the flame quickly. The slide should not remain in the flame so long that it becomes too hot to handle.
6. Place the slide on a paper towel and apply the stain to the dried material on the slide. Allow the stain to remain on the slide 3-5 minutes. Handle the stain carefully because it will do just what its name implies—stain. Bacterial stains will eventually come off the skin but may stain clothes permanently.
7. Now use a plastic squeeze bottle of water or the flow of water from a tap to gently wash the stain off the slide.
8. Use a paper towel to ***blot***, not rub, the slide dry.
9. Your slide is now ready for viewing, so put it on the microscope and prepare to enter the hidden world of microbes.

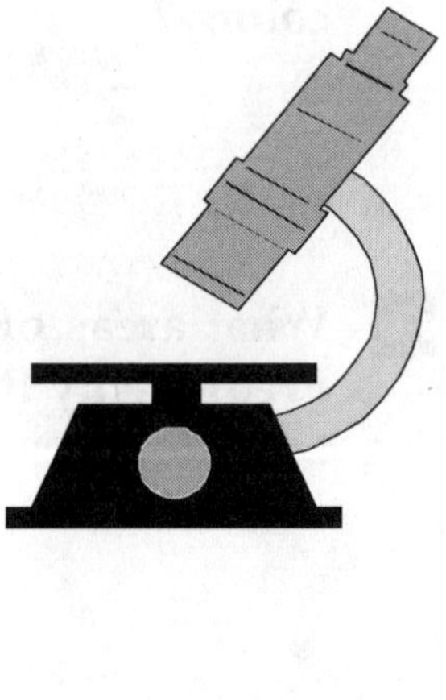

Observing Bacteria

Continued

1 Why is it important that your slide be very clean?

2 What are bacterial stains and why do bacteriologists use them?

3 Why did you dry your slide after the liquid bacteria culture was placed on it?

4 Why was the stain left on the dried bacteria for several minutes?

5 Why were you instructed to blot, not rub, the slide dry?

6 Diagram and describe the bacteria you saw.

Build a Bacterial 'City'

In this activity, we will visit the hidden bacterial world found in muds and sediments by building a soil column device.

The bacterial "city" you will construct is similar to Earth: the energy source is the same as ours, it possesses oxygen in the atmosphere, and it is composed of soil and mud. One difference is that the surface of our "city" is completely flat and flooded by a thin film of water.

The "city" consists of translucent "buildings" (layers) populated by microbial life forms, living soil, and water. These are the same life forms that exist in flooded environments on our own planet—pond bottoms, salt marshes, and flooded agricultural fields.

Your "city" will appear lifeless at first, but be patient. With an influx of energy and some simple management, the spectacular inhabitants of your bacterial "city" will show themselves.

Procedure

1. To start your microbial "city" you will need a translucent (allows light through) clear container, which your teacher will supply.

2. Use a bottle top as a funnel to fill your container with the mud, soil, and/or gunk that you, your group, or your teacher has collected.

3. Pour a few centimeters of mud through the funnel; then, with one hand covering the top opening of the bottle and the other hand holding the bottle on the side, strike the base of the bottle firmly on the table once to settle the mixture evenly.

4. Continue filling the bottle, hitting the base on a table every few centimeters until the bottle is filled to within 4 or 5 centimeters of the top. Allow the contents of the bottle to settle for 24 hours so there is about 1 cm of clear water on the surface. If there is too much or too little water, add or remove some with a dropper.

5. Now cover your "city" with clear plastic wrap fastened with a rubber band. Write the names or initials of each group member and the date on a small piece of masking tape and fasten it to the bottom of the bottle.

6. Place your bottle in a well-lit place, such as a bright window that does not get too much light. Even better, set up an incandescent light(s) of 40–60 watts illuminating the top and sides (but placed far enough away so as not to heat the bottle). You can keep the lights on continuously or put them on a 12-hour timer.

7. Measure layers of your city regularly and record observations on the observation sheet. Show different layers on the column.

Build a Bacterial 'City'

Continued

Environmental Factors in the 'City'

The bacteria live at different levels in the "city" because the environmental conditions at each level are unique. The following elements help determine at which level specific bacteria live:

- **Oxygen**—It dissolves into the water from air at the top of the "city." As we go deeper down to the lower levels, the amount of oxygen decreases until at the bottom there is none. In fact, going down the column is like going back in time! For millions of years, no oxygen was in the atmosphere, and the bacteria that lived then developed without it.

- **Sulfur**—It is released after living organisms die, as bacteria and fungi decompose the tissue and body parts. Often, these dead parts are buried, so the sulfur is released from below and diffuses upward. Therefore, the concentration of sulfur in your "city" increases as you go deeper. Sulfur is released in the form of hydrogen sulfide, which has the familiar smell of rotten eggs. You may have smelled this while walking along the edge of a pond, stream, or mudflat.

- **Photosynthesis**—Colored layers should develop in your "city" in three to six weeks. These colored layers are bacteria undergoing photosynthesis (making food using light). The greenish layers on top use the wavelengths of visible light for photosynthesis while the lower layers use infrared light.

For Further Investigation

1. Devise a way to remove some of the bacteria from each level so you can observe them under the microscope.

2. What effect does covering a portion of the "city" with black construction paper have? Compare covered areas with uncovered areas.

3. What effect does wrapping the "city" in colored cellophane have? Test different colors.

Build a Bacterial 'City'

Continued

OBSERVING A SOIL COLUMN

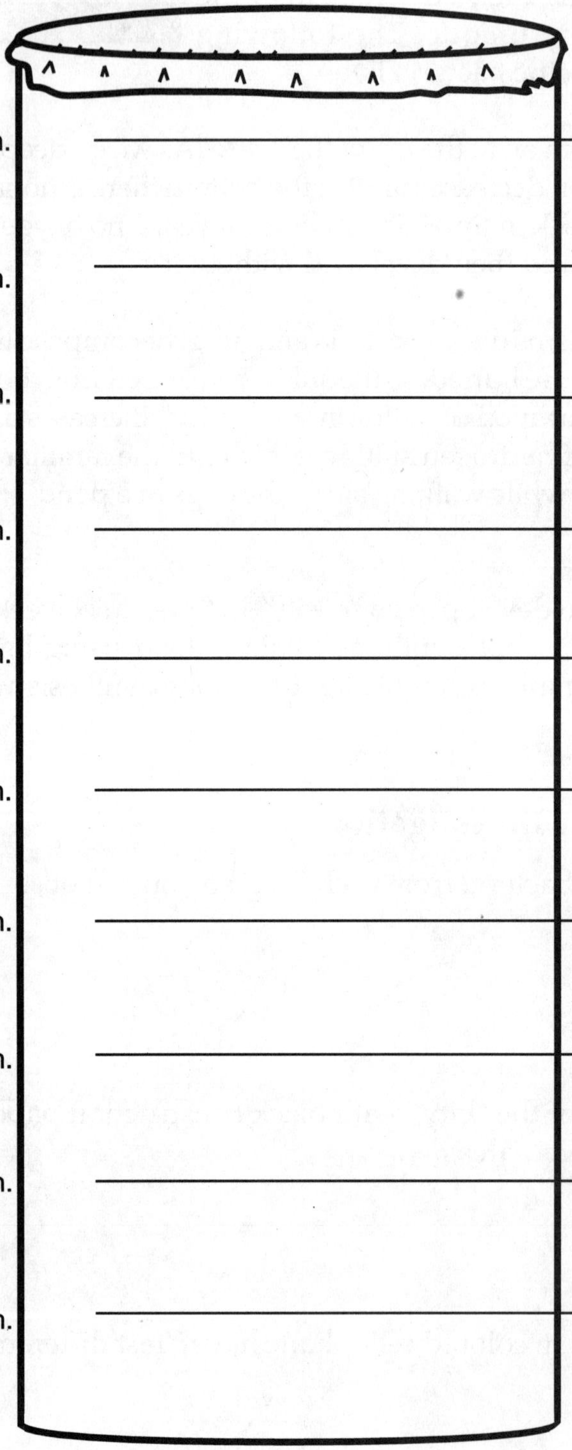

cm. mark	GRADIENT COLOR	OBSERVATIONS
20 cm.		
18 cm.		
16 cm.		
14 cm.		
12 cm.		
10 cm.		
8 cm.		
6 cm.		
4 cm.		
2 cm.		

Make Your Own Yogurt

Humans cleverly harness bacteria for many useful purposes. In this activity you will investigate our dependency on microbial metabolism for the manufacture of dairy products by producing MYO (Make Your Own) yogurt. During the process, prepare to be amazed at how fast bacteria can reproduce and multiply.

Procedure

1. Open two cans of condensed milk and pour out half of each.
2. Fill the remainder of each can totally with water that is distilled or that has been boiled for 20 minutes.
3. To one of the cans add a dropper of active cultured plain yogurt. This is your **experimental** (test) can. Do not add anything to the second can. This is your **control** (comparison) can.
4. Lightly cover both cans and place them in a warm place.
5. Observe and describe what is happening in both cans.
6. After two or three days, you will be able to eat your yogurt.

For Further Investigation

1. After several days, use a dropper to remove some of the liquid from the can containing the yogurt culture (the test can), prepare and stain a slide of this liquid, and observe through the microscope.
2. Let one test can and one control can sit for approximately 10 days. Explain any differences you see between the two cans.

A War On Bacteria

Killing or at least containing bacteria has been a human goal ever since the connection between bacteria and disease, death, and decay was first realized.

In this activity you will attempt to determine the effectiveness of some of the chemicals that have been developed to kill bacteria.

Procedure

Part I: Antibiotics

By lucky accident in 1929, Dr. Alexander Fleming discovered the bacteria-killing properties of antibiotics. Dr. Howard Florey perfected the use of antibiotics as a medical tool to kill bacteria. Most antibiotics are purified products from molds. Incredibly, antibiotics kill bacteria but not other types of cells, making them ideal for internal (inside) use in humans and animals.

1. Get one dish of nutrient agar per group.
2. Use a marker to put your initials or another identifying symbol on the lid.
3. Divide the bottom of the dish into the number of sections called for by the teacher.
4. The teacher will now swab bacteria on the surface of the agar in your dish.
   - Some groups will get *Escherichia coli*—a common inhabitant of the human intestinal tract that does not form spores.
   - Some groups will get *Bacillus cereus*—a common rod-shaped decay bacteria that does form spores.
   - Record the type of bacteria your group received under question 1 in the Think About It—Part I section.
5. Now add paper discs that have been soaked in a variety of antibiotics to the surface of your agar. Open the lid of your dish, place the paper disc on its section with a pair of forceps, lightly touch the paper disc to stick it to the surface of the agar and then close the lid. Quickly replace the lid after adding each disc. Place only one paper disc in each section of the dish.
6. Record the name of each type of antibiotic we tested under question 2 in the Think About It—Part I section.
7. ***Safety Notes***—Your dish contains large numbers of bacteria so:

A. Replace the lid quickly after adding each disc.

B. Do not touch the agar with your fingers, pencils or anything other than the forceps.

C. Be careful not to lay the forceps down on the table between adding discs or touch anyone or anything with the forceps while working in the dish. The forceps contain large amounts of bacteria on the tips.

D. Once you have added all the discs to your dish, place the forceps tip down into a container of bleach and thoroughly wash your hands.

A War On Bacteria

Continued

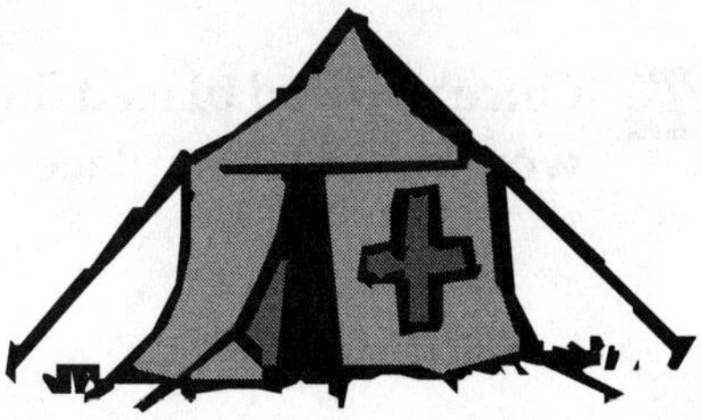

E. Keep your hands away from your face and eyes at all times and thoroughly wash your hands when you are done.

8. Place your dish upside down in a warm place (preferably an incubator) for 24 to 48 hours then examine and answer the questions that follow.

Part I: Think About It

1 **Which type of bacteria did your group receive?**

2 **What types of antibiotics did we test?**

3 **Why did the teacher cover the surface of the agar with bacteria?**

4 **Why were you to put your plates somewhere warm?**

A War On Bacteria

Continued

5 **Once you had placed the antibiotic papers in your dish, why were you told to place your dish upside down?**

6 **What safety issues are we dealing with in this activity?**

7 **Did the antibiotics work? How can you tell?**

8 **Which antibiotic worked the best? How can you tell?**

9 **Did each type of antibiotic work equally well against each type of bacteria? Explain.**

10 **Which type of bacteria should have been the hardest to kill and why?**

A War On Bacteria

Procedure

Part II: Common Household Chemicals

Humans have also developed an array of chemicals to kill bacteria on countertops, in your toilet, and even on your teeth. In this part of the activity, we will test the effectiveness of some of these chemicals.

1. Look at the selection of chemicals gathered by the teacher and/or brought by students.

2. Select the chemicals your group wishes to test. The exact number of chemicals each group can work with will be determined by the teacher.

3. Get one dish of nutrient agar. Using a marker, put your initials or some type of identifying mark on the lid and divide the bottom into sections that correspond to the number of chemicals you are going to test.

4. The teacher will now swab bacteria on your plate. We will use the same types of bacteria that we did in Part I with half the groups getting one type and half the groups the other. Record which type of bacteria the teacher puts in your dish under question 1 of the section Think About It—Part II. If you are going to test a germ-killing mouthwash, inform your teacher. The teacher will not put bacteria on the section where the mouthwash disc will be. Instead, take a clean cotton swab, vigorously rub your teeth, then swab this onto the section of the agar where the mouthwash disc will go.

5. *Safety Notes*— Your dish contains large numbers of bacteria, so
   A. Replace the lid quickly after adding each disc.
   B. Do not touch the agar with your fingers, pencils or anything other than the forceps.
   C. Be careful not to lay the forceps down on the table between adding discs or touch anyone or anything with the forceps while working in the dish. The forceps contain large amounts of bacteria on the tips.
   D. Once you have added all the discs to your dish, place the forceps tip down into a container of bleach and thoroughly wash your hands.
   E. Keep your hands away from your face and eyes at all times and thoroughly wash your hands when you are done.

6. Use the forceps by each sample of chemical to pick up a paper disc. Holding the disc with the forceps, soak the paper in the chemical you wish to test. Carefully wipe the disc on the edge of the sample jar several times.

7. Open your dish and carefully place the paper disc soaked in chemical in its proper section. Use the forceps to lightly touch the disc to stick it down to the agar.

8. Replace the lid and repeat. Use separate forceps for each chemical sample; do not mix up the forceps.

9. Once you have all chemicals to be tested added to your dish, place your dish upside down in a warm place (preferably an incubator) for 24 to 48 hours, then examine and answer the questions that follow.

A War On Bacteria

Continued

Part II: Think About It

1 Which type of bacteria did the teacher put in your dish?

2 Section by section, what chemicals did your group test?

3 If you tested mouthwash, why did you swab that section with material from your teeth/ mouth instead of the bacteria used on the other parts of the dish?

4 Did some chemicals work against one type of bacteria but not against the other type? Explain.

5 Which chemicals seemed to work the best overall?

6 In conclusion, would you say that these chemicals are as effective as they are advertised to be? Defend your answer.

Putting It On Paper

A. A Formal Library Report

Write a report on how bacteria affect the things around them. You may take a specific approach and discuss how bacteria affect just humans or take a more general approach and explain how bacteria interact with all living things.

B. A Day in the Life of a Bacterium

Imagine you are a bacterium—you could be a decay bacterium in soil, a cyanobacterium in pond water, or a disease-causing bacterium inside a human body. Write a short story in which you describe your existence for one day. In your story, explain what you look like, what dangers you face, how you will get food, and how you reproduce.

and/or

Imagine you are a bacterium that can see. Paint a picture of what your habitat looks like around you. Be prepared to explain your picture to the rest of the class.

C. The Great Debate

Imagine the time is the distant future. In the Great Hall of Biology, a debate is raging. One side wants to eliminate microbes from Earth; the other side wants to keep them. Which side would you choose and why? Defend your position in writing with facts.

Answers

Bacteria Concept Map

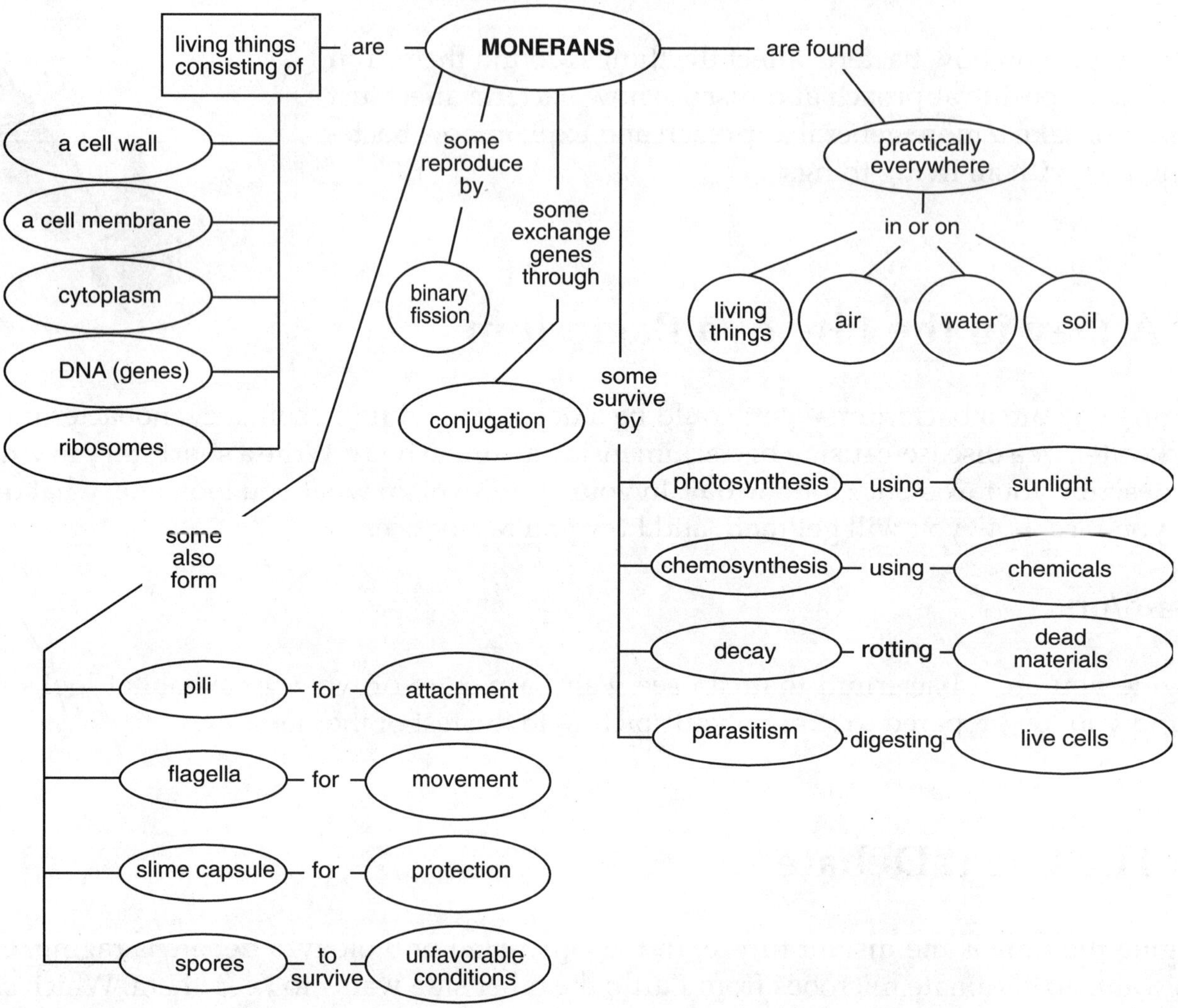

Answers

Bacteria Brain Bogglers

1. at the end of 1 hour = 8 bacteria
   6 hours = 262,144
   12 hours = 68,719,476,736

   To make the point of rapid increase another way, have students imagine this scenario: A student has suggested to her parents that she receive an allowance of one penny a day with the amount doubled each day. Her parents quickly agree. Who was the more clever?

   This young lady will make 1 cent on the first day, 2 cents on the second day, 4 cents on the third day and so on. Doesn't sound like much, right? However, on day 31, her parents would owe her $10,737,418.24 (yes, dollars, not total pennies). Fortunately, bacteria cannot keep up this type of rapid reproduction very long or a layer of bacteria covering the entire planet many feet thick would have settled in long ago.

2. Stage 1: Population is rapidly increasing.
   Stage 2: Population is stabilizing.
   Stage 3: Population is declining.

   The leveling off in Stage 2 is due to the fact that you have as many bacteria as the habitat can support; in Stage 3, they begin to decline because you have more bacteria than the habitat can support.

3. Patient 1 had salmonella food poisoning; Patient 2 had staphylococci food poisoning. The former takes longer to run its course.

A Bacteria 'Trap'

1. The nutrient agar serves as a food source for the bacteria.
2. The sticky side of the tape picks up bacteria from surfaces or objects and allows us to transfer them to the agar.
3. The purpose of touching Section 1 in the dish is to transfer skin bacteria so we might observe their growth.

A Bacteria 'Trap' (cont.)

4. The lids of the dishes should be kept on for safety reasons. There is always a chance, slim though it is, that a pathenogenic (disease-causing) strain of bacteria may have been transferred and managed to grow in the dish. Also, even harmless bacteria growing condensed in large numbers in a lab situation could be potentially harmful. Leave the lids on and be better safe than sorry.
5. The dishes were placed upside down to keep water from condensing on the lid and dropping onto the agar surface. Bacteria will grow in a "lake" of water on the surface of the agar, but you will not be able to see individual colonies.
6. A bacterial colony is a visible growth of perhaps tens of millions of bacteria.
7. Answers will vary with location.

Observing Bacteria
(Suggested Answers)

1. Because of the extremely small size of bacteria, smudges and dust may obscure them if slides (and microscope lenses) are not thoroughly cleaned.
2. Bacterial stains are special dyes that bacteriologists use to make bacteria more visible. The students should have been able to see the bacteria much more easily after they were stained.
3. The slide is dried for two reasons: (1) to kill the bacteria, rendering them harmless, and (2) to stick bacteria to the slide so they are not washed off when the stain is removed.
4. The stain must be left on long enough for it to penetrate the bacteria's tough cell wall.
5. If students rub the slide dry, they will wipe off the dried bacteria.
6. Answers will vary.

Answers

Build a Bacterial 'City'

1. You can sample different layers in the bottle by puncturing the side with the sharp point of a mathematician's compass, removing a sample with a dropper, and examining it under a microscope. To reseal the column, use silicon sealant. Puncturing the column can release strong odors, so you may want to do this outside.
2. Answers will vary.
3. Answers will vary.

Make Your Own Yogurt

1. If students make and observe slides of the bacteria from both cans, they will note many more bacteria in the test (yogurt) can than in the control (uninoculated) can. The few bacteria in the control can got into the milk from the air.
2. After about ten days, mold will begin to grow across the surface of an uninoculated (control) can but not in a yogurt (test) can. Have students offer explanations as to why only the control can molds. The reason is that yogurt bacteria feed on the sugars of the milk (lactose), turning them to lactic acid. Molds are inhibited by the lactic acid, plus they find little if any food (sugars) available since the bacteria used it up.

A War on Bacteria

Part I

1. Students should write the specific type of bacteria they received in their dish.
2. Students should list the antibiotics tested and the section of the dish containing each.
3. Swabbing on bacteria produces a bacterial "lawn" completely covering the surface of the agar. If the bacteria around the discs are killed, a clear zone will appear in this whitish bacteria growth.

A War on Bacteria (cont.)

4. Both types of bacteria used in this activity grow best from 30° to 37°C. They will grow at room temperature, only more slowly.
5. Bacteria plates are incubated upside down to prevent moisture from the lid dropping down on the surface of the agar and turning our bacterial "lawn" into a bacterial "lake" that would obscure results.
6. See Safety Notes in the student procedure.
7. If the antibiotics work, students will see a clear zone around the disc.
8. The bigger the clear zone, the more bacteria that were killed by the antibiotic.
9. The B cereus should be harder to kill (smaller clear zones) than the E. coli. B cereus forms spores that are somewhat resistant to antibiotics; E. coli does not form spores and is easier to kill.
10. See answer for 9.

Part II

1. Students should record the specific type of bacteria in their dish.
2. Students should record section by section which chemicals they tested.
3. It is not a fair test if we don't use the type of bacteria the mouthwash is made to kill.

4–6. Some of the chemicals tested will work quite well (large circles), and others will not. As in Part I, the spore-forming B. cereus should be the hardest for the chemicals to kill (smallest circles). Caution students not to jump to unjustified conclusions. If some chemicals do not work as effectively as advertised, it may be due to the crudeness of our test design and not a fault with the chemical(s).

Classifying Life on Earth

THE FIVE KINGDOMS

ANIMALIA

PLANTAE

MONERA

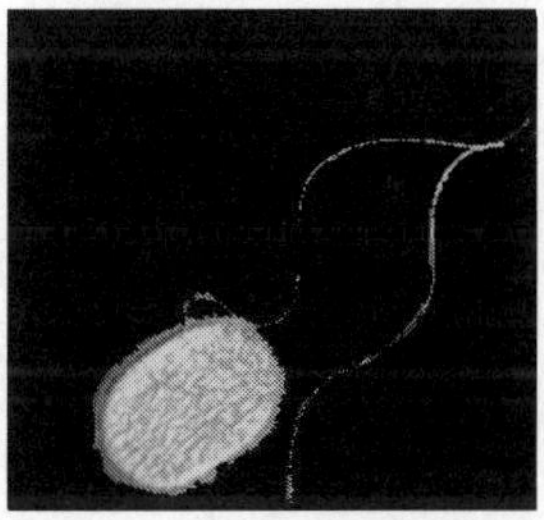

FUNGI

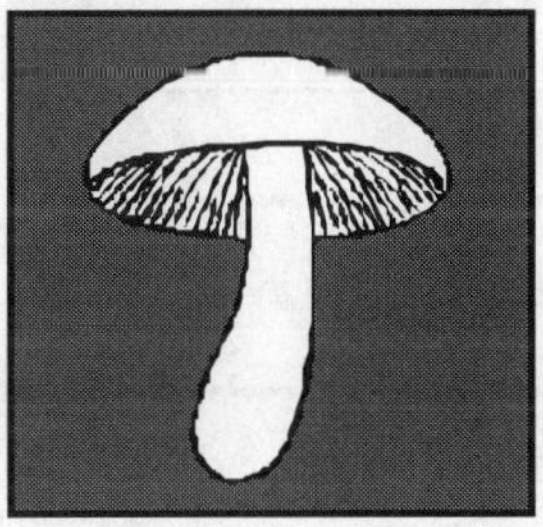

PROTISTA

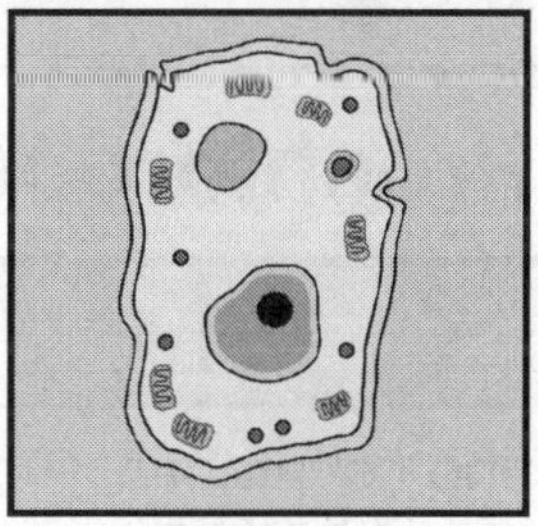

THE FOUR PHYLA OF KINGDOM MONERA

Archaebacteria

Methanogens
found at bottom of bogs, swamps, sewage treatment plants

Halophiles
found in high-salt areas, such as Dead Sea, Great Salt Lake

Thermoacidophiles
found in boiling, acidic hot springs

Schizophyta

Eubacteria
free-living in soil and water

Actinomycota
can cause diseases such as tuberculosis

Rickettsiae
can reproduce only in the cells of a specific host

Spirochaeta
can cause Lyme disease, syphilis

Prochlorophyta

photosynthetic, live within bodies of tunicates—marine organisms

Cyanophyta

photosynthetic, producing own food, found primarily in water

Small but plentiful ...

1 ounce = 30 trillion bacteria

Structure of a Bacterial Cell

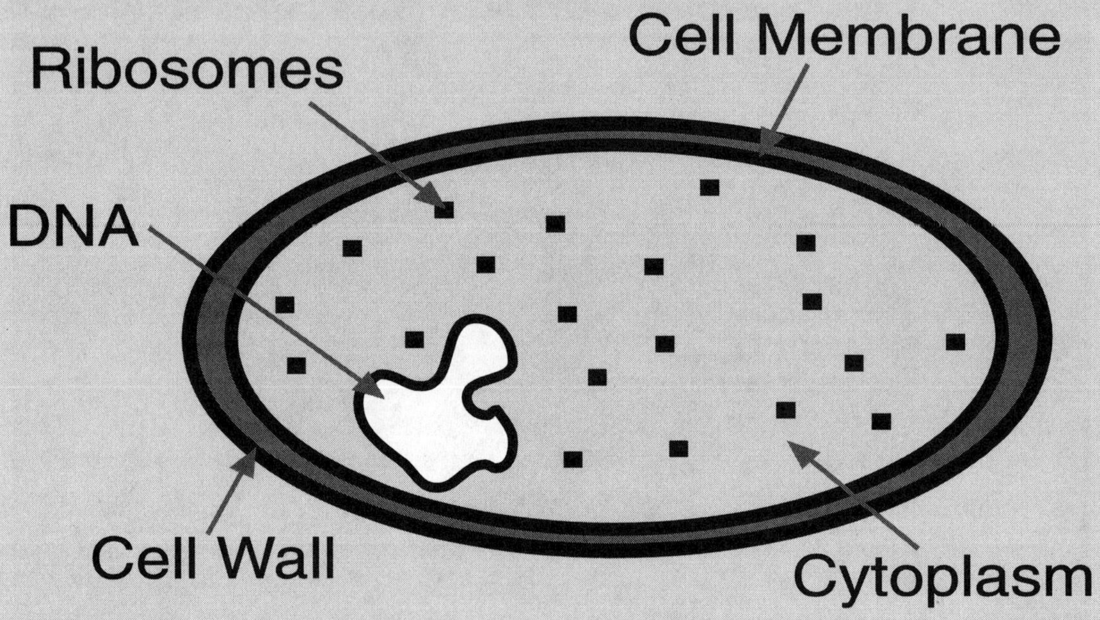

Secondary Structures of Bacteria Cells

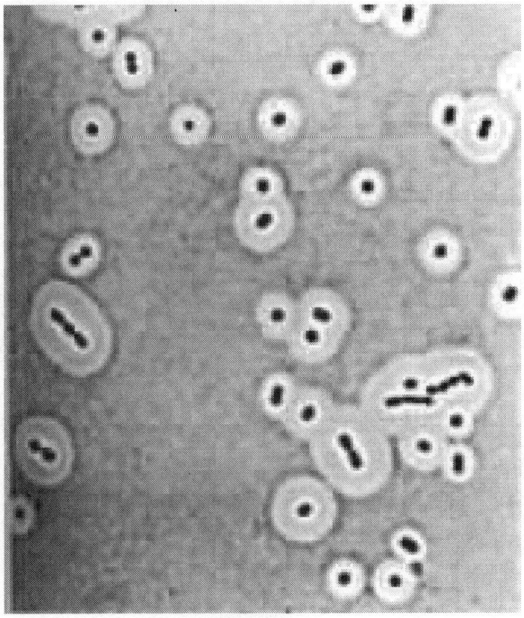

Capsules

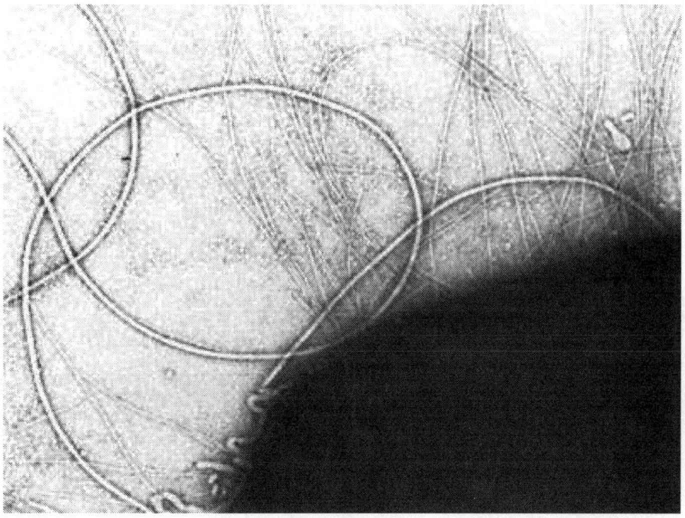

Pili

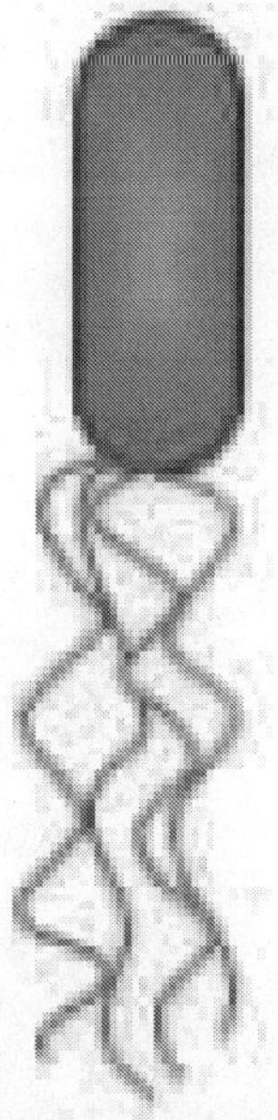

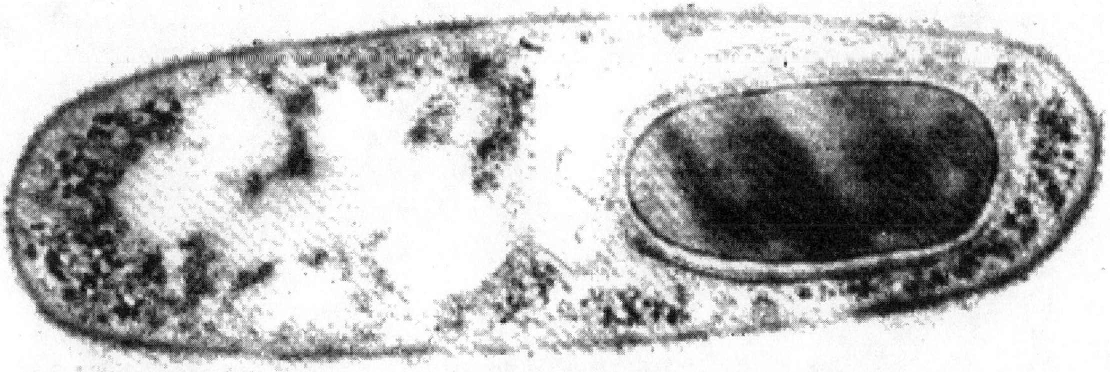

Endospore

Flagella

Chemosynthetic Bacteria Recipes

$CO_2 + H_2$ = Food

$CO_2 + NH_3$ = Food

Carbon dioxide mixes with hydrogen gas or ammonia to produce food.

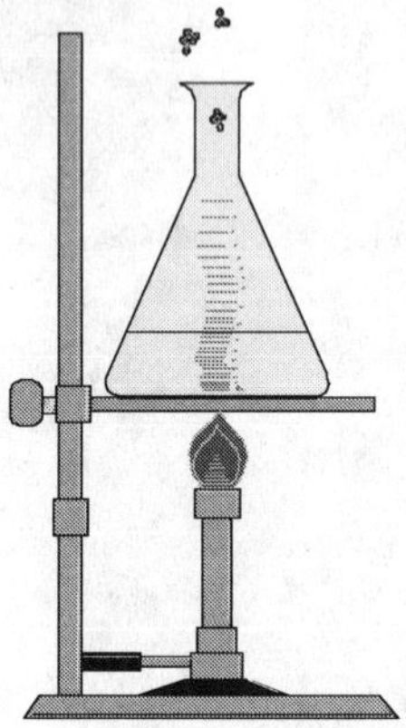

TWO GROUPS OF MONERANS

Autotrophs

produce their own food through photosynthesis (from chlorophyll) or chemosynthesis (from carbon dioxide)

Heterotrophs

get food from their environment (by feeding on dead remains)

Appendix

The companies listed here can supply the materials—bacteria cultures, glassware, media, slides, microscopes—you need to conduct the activities outlined in this book. Each company's technical support division can also provide valuable information and suggestions when you have a problem or question.

Carolina Biological Supply Co.
2700 York Road
Burlington, NC 27215
Phone: 1-800-334-5551

Catalog

A Biology Materials catalog is sent free to schools if requested on school stationery.

Shipping Requirements

Recognized educational institutions and full-time teachers may order what they wish with no minimum dollar value involved; terms are 30 days net from the date of the invoice. Individuals must prepay by check, money order, or credit card, with a minimum order of $15. Orders will not be accepted by individuals for chemicals, kits containing chemicals, or liquid-preserved organisms.

Materials Available

Microscopes, live bacteria cultures, books, charts, media, slides and other glassware, stains, and assorted kits involving various bacteriological concepts.

Connecticut Valley Biological Supply Co.
Box 326
82 Valley Road
Southampton, MA 01073
Phone: 1-800-628-7748

Catalog

Connecticut Valley will supply catalogs free of charge upon request by school officials, teachers, or private individuals.

Shipping Requirements

No minimum purchase is required of schools, but orders from schools must be accompanied by a purchase order number. Terms are net 30 days. The minimum order for individuals is $10 and these orders must be accompanied by a check, money order, or credit card authorization for the full amount. Connecticut Valley will not sell chemicals to individuals.

Materials Available

Microscopes, live bacteria cultures, books, charts, media, slides and other glassware, stains, and assorted kits involving various bacteriological concepts.

Ward's
P.O. Box 92912
Rochester, New York 14692-9012
Phone: 1-800-962-2660

Catalog

Ward's will supply catalogs free of charge upon request by school officials or teachers.

Shipping Requirements

Credit is extended to all educational institutions. Terms are net 30 days from the date of the invoice. Teachers' personal orders are welcome.

Materials Available

Microscopes, live bacteria cultures, books, charts, media, slides and other glassware, stains, and assorted kits involving various bacteriological concepts.

The listing of the suppliers in this appendix does not represent an endorsement of those companies or a guarantee of their products or services.

